Guide to Non-Ferrous Metals and their Markets

Guide to Non-Ferrous Metals and their Markets

Peter Robbins

Kogan Page, London/Nichols Publishing
Company, New York

The author would like to thank Kristine Hatch, Anita Kogan, Stephen Moss, David Young, Sandy Young and Ian Youngs for their help in the preparation of the third edition of this book, and the Metal Bulletin for providing production statistics.

First published in Great Britain in 1979 by
Kogan Page Ltd, 120 Pentonville Road, London N1 9JN
Second edition published in 1980
Third edition published in 1982

British Library Cataloguing in Publication Data

Robbins, Peter
Guide to non-ferrous metals and their markets.
3rd ed.
1. Nonferrous metal industries
I. Title
338.4'7669 HD9539.Al

ISBN 0-85038-524-5

Published in the United States of America in 1979, 1980, 1982
by Nichols Publishing Company, Post Office Box 96,
New York, NY 10024

Library of Congress Cataloging in Publication Data

Robbins, Peter.
Guide to non-ferrous metals and their markets.
Rev ed of: Guide to non-ferrous metals and their markets / Peter Robbins. 1979.
1. Nonferrous metal industries. 2. Market surveys.
I. Guide to non-ferrous metals and their markets.
II. Title.
HD9539.A2R63 1982 380.1'45669 81-18897
ISBN 0-89397-124-3 AACR2

Printed in Great Britain by the Anchor Press Ltd
and bound by William Brendon & Son, both of Tiptree, Essex

Contents

Part 1: Production, Development and Marketing

Introduction 11
Supply Disruption and the Resource War 13
Future Supplies 15
Speculation and Investment 19
Pricing of Non-Ferrous Metals 25
Substitution 29
How Metals are Traded 31
Summary of the Non-Ferrous Metals Market 35
The Role of Non-Ferrous Metals 39

Part 2: The Metals

Aluminium 45
Antimony 51
Arsenic 55
Beryllium 58
Bismuth 61
Cadmium 65
Cerium 69
Chromium 72
Cobalt 76
Columbium or Niobium 79
Copper 82

Gallium 87
Germanium 90
Indium 93
Lead 96
Lithium 100
Magnesium 103
Manganese 106
Mercury 109
Molybdenum 112
Nickel 115
Rhenium 119
Selenium 122
Silicon 125
Sodium 129
Tantalum 131
Tellurium 134
Tin 137
Titanium 141
Tungsten 144
Vanadium 149
Zinc 152
Zirconium 156

Appendices

Appendix 1: Useful Addresses 161
Appendix 2: Glossary of relevant terms 169
Appendix 3: Average monthly dollar/sterling rate based on the market price at 11.30 am each dealing day 173
Appendix 4: World Time Zones 175
Appendix 5: London Metal Exchange contract rules and arbitration 177

Appendix 6: Conversion table 185

Appendix 7: Conversion of US metal prices from cents per lb to sterling pounds per metric ton at varying exchange rates 187

Appendix 8: Malaysian tin prices in ringgits per pikul to sterling pounds per metric ton 189

List of diagrams and price graphs

Diagrams

Aluminium metal production, 1980 45
Aluminium metal consumption, 1980 46
Antimony metal production, 1979 51
Arsenic (trioxide) production, 1979 55
Beryllium ore production (metal content), 1977 58
Bismuth metal production, 1979 61
Cadmium metal production, 1979 65
Cadmium metal consumption, 1979 66
Cerium (monazite ore) production, 1975 69
Chrome ore production, 1979 72
Cobalt metal production, 1978 76
Columbium ore production, 1979 79
Copper refined production, 1980 82
Copper refined consumption, 1980 83
Gallium production, 1975 87
Germanium ore production, 1978 90
Indium metal production, 1980 93
Lead refined production, 1980 96
Lead refined consumption, 1980 97
Lithium ore production, 1977 100
Magnesium metal production, 1979 103
Manganese ore production, 1979 106
Mercury metal production, 1979 109
Molybdenum ore production, 1979 112
Nickel metal production, 1980 115
Nickel metal consumption, 1980 116
Rhenium metal production, 1978 119
Selenium metal production, 1979 122
Silicon metal production, 1978 125
Sodium metal production, 1978 129
Tantalum ore production, 1979 131
Tellurium metal production, 1979 134
Tin refined production, 1980 137
Tin refined consumption, 1980 138

Titanium sponge estimated production, 1980 141
Tungsten ore production, 1978 144
Vanadium ore production, 1979 149
Zinc metal production, 1980 152
Zinc metal consumption, 1980 153
Zirconium ore production, 1979 156

Price graphs

Aluminium metal prices (free market LME cash settlement) 50
Antimony metal prices (free market) 53
Arsenic metal prices (producer price) 56
Beryllium ore prices (producer price) 59
Bismuth metal prices (free market) 63
Cadmium metal prices (free market) 67
Cerium (mischmetal) prices (UK producer prices) 71
Low carbon ferro-chromium prices (free market) 74
Cobalt metal prices (free market) 78
Columbium ore prices 81
Copper prices (LME cash settlement wirebar) 85
Gallium metal prices (free market) 88
Germanium metal prices (30 ohm) (free market) 92
Indium metal prices (free market) 95
Lead prices (LME cash settlement) 98
Lithium metal prices (producer price) 102
Magnesium metal prices (free market) 105
Manganese ore prices (free market) 108
Mercury prices (free market) 111
Molybdenum ore prices (producer price) 114
Nickel metal prices (free market) 118
Rhenium metal prices (free market) 121
Selenium metal prices (free market) 124
Silicon metal prices (free market) 127
Tantalum ore prices (free market) 132
Tellurium metal prices (producer price) 135
Tin prices (LME standard tin-cash settlement) 140
Titanium (sponge) metal prices (free market) 142
Tungsten ore prices (free market) 147
Ferro-Vanadium prices (free market) 150
Zinc prices (LME cash settlement) 155
Zirconium ore prices (producer price) 157

Part 1: Production, Development and Marketing

Introduction

A great deal of public attention has been focused on non-ferrous metals in recent years. This has occurred because of a sudden realization that industries in the Western world are dependent on Third World countries for much of their essential supplies. The origin of this concern can be traced to a number of comparatively recent events.

First, the success of oil-exporting countries (represented by OPEC) in raising oil prices at a much faster rate than average world inflation demonstrated, for the first time, that suppliers acting in concert are able to deny industrial countries cheap raw materials.

Second, the boom in electronics and other sophisticated areas of technology has made the Western world more dependent on supplies of some obscure metals, many of which are only to be found in commercially viable deposits in unstable Third World countries. The rebel invasion of Sharba (the world's major source of cobalt) in 1978 showed just how vulnerable to disruption mineral supplies can be.

Third, the huge surge in demand for most metals, coupled with roaring inflation in 1978 and 1979, caused prices to escalate to historically high levels. This boom coincided with near-panic buying by investors of precious metals.

Concern has been exacerbated by a theory held by many influential people, especially in the United States, that the West is now engaged in a 'resource war' with the Soviet Union, which is supposed to have a strategy of undermining Western economies and defence capabilities by denying it its raw materials. Just what grounds there are for such concern, however, is a matter of conjecture.

1981 was marked by a fall in non-ferrous metal prices as dramatic as the boom in prices in the years that preceded it. The fall was so great that many metals are now being produced at below the cost of production. Metals like cadmium, bismuth and selenium, for example, are cheaper now than they ever have been. Efforts by CIPEC (the copper producers' equivalent of OPEC) and the IBA (the bauxite producers' association) have failed to keep prices at their previous levels, let alone increase them. The prospects of cartelization of metal markets seems to be as distant a likelihood as ever. Most OPEC countries share the same religion, language and geographical location, facts which seem to have played a great part in the success of the organization, but no such advantages apply among metal producers and the effectiveness of their joint action has been correspondingly less.

Western industry is becoming more dependent upon high technology, especially in electronics, and industries in the West cannot be sustained without supplies of certain rare metals. A critical shortage of supplies of a number of metals occurred in the late 1970s. The price of tantalum, used to make electronic capacitors, rose by 600 per cent between 1977 and 1980. Titanium prices increased by 500 per cent in the same period. Titanium's major use is in the manufacture of civil and military aircraft and missiles, and the USSR is said to have built a new class of submarine with a titanium hull. Molybdenum, a metal much favoured as an alloy ingredient for making oil pipelines, has a similar recent price history and the same can be said of indium, germanium and the platinum group metals. These last three have numerous uses in advanced electronic systems.

It should be noted, however, that the prices of all these metals (with the exception of germanium) have since fallen back to levels similar to those before the boom started. These price falls can, in part, be blamed on the recession, but technological advance has not slowed down. Advanced electronic systems have been employed everywhere from wrist watches to the supermarket and a great deal more money is being spent on armaments, an area with a very high intake of obscure metals.

Several factors have been at work to reduce prices. High prices have naturally encouraged increased production. Titanium production in Japan, for instance, has grown from 7,000 tons in 1977 to 20,000 tons in 1981. The recovery of scrap metal has been made more efficient and substitutes have been found for certain high-priced metals for some uses.

Supply Disruption and the Resource War

In May 1978 a group of Katanguese rebels crossed the border from Angola, where they had been living as refugees since the Congolese civil war of the 1960s, into the mineral-rich province of Sharba (formerly Katanga) in Zaire. They killed a number of white mining engineers and their families and committed various acts of sabotage in the mines, before being driven out by French and Belgian troops.

At that time Sharba was responsible for producing approximately half the world's cobalt, which was produced as a by-product in the refining of the cobalt-rich copper ore, found in the area. Damage to the mines was fairly superficial but a significant number of white expatriate engineers involved in mining and refining decided to leave the country. The price of the remaining supplies of cobalt climbed by 400 per cent after the raid.

This single event forms the keystone of a theory which has fundamentally changed the Western world's attitude to its supplies of non-ferrous metals, because the raid was supposed to have been planned and orchestrated by the USSR. According to the theory, the USSR is engaged in a strategy of disrupting supplies of strategically important minerals to Western industrialized countries. This, it is suggested, is achieved by supporting revolutionary movements in the Third World countries where these supplies are generated.

Many of the materials encompassed by this theory have indispensable uses in the armaments industry — metals such as germanium which is used to make electronic nightsighting instruments and tungsten which is used to make shells and bullets. By disrupting supplies of these materials to the West the USSR could theoretically prevent sophisticated armaments being produced in the West. It is extremely difficult, however, to square this theory with events.

In 1970 INCO, the giant Canadian nickel producing company, went on strike. At that time INCO produced about one-third of the world's nickel and supplied the USA with about three-quarters of its requirements. The strike caused the price of nickel to rise by over 300 per cent. But is was the sale of huge quantities of Soviet nickel to the Western market at that time that prevented the price from going much higher and stopped the West from running out of metal. Much of the nickel sold by the USSR is believed to have come straight from its own strategic stockpile.

The USSR is almost self-sufficient in its requirements for most metals and is a large exporter of many of them. It remains, in fact, one of the largest suppliers of strategically important metals to the USA. The revenues from these sales represent a large proportion of much-needed hard currency for the USSR.

It may be worth examining the 1978 Sharba raid a little more closely because it is the one event on which the 'resource war' theory hangs.

The invading rebels were Katanguese. The reason they were forced to leave the Congo in the 1960s was that they associated themselves with the movement to separate Katanga from the rest of the country. This movement was, according to G Lanning and M Mueller, writing in *Africa Undermined*, backed by several Western multinational mining companies and the Belgian Government. There is some evidence for their assertion. The rebels' original battle was *against* the Government of Patrice Lumumba, a pro-Soviet Marxist. Moreover, they were able to use part of the Benguela railway, which is controlled by UNITA, the South African backed Angolan rebel movement, to move heavy equipment from Angola into the invaded area. Whether or not the rebels were backed by the West, no evidence has been found to link the invasion with either Cuba or the USSR.

The 'resource war' theory is just a theory. The political allegiance of some Third World countries swings back and forth between pro-West and pro-East regularly, but efforts by those countries to sell metal to the West remains constant because, in most cases, revenue from the sale of those metals is needed to survive.

Nevertheless, the theory has enabled certain political elements in the West, especially in the USA, to justify a radical overhaul of its minerals policy. These changes include efforts to stimulate production of metals in the West, which has, for example, meant that some American national parks are no longer protected from exploitation by mining companies; a much less generous attitude on the part of the West towards Third World countries in discussions on the law of the sea; the building of strategic stockpiles in Japan, France, Sweden, and the USA, and a more tolerant stance toward South Africa, a major producer of many strategically important metals.

Future Supplies

It is generally true to say that as rich ores in accessible areas become worked out, less rich ores in inaccessible areas have to be exploited. It is also a fact that the cost of mining and refining metals is increasing. These facts, however, need to be qualified.

The technology used to prospect and produce metals is constantly improving. Grades of ore that would not have been considered workable a decade ago can now be treated. This is the result not only of the higher prices obtainable for the product but of the more efficient extraction techniques used.

Although most of the earth's surface has been subjected to geological surveys (with the exception of Antarctica, and some parts of Siberia, central Africa and South America), infra-red aerial photography, improved seismology and computerized data analysis have aided the discovery of new deposits of metal ore in remote areas. Such improvements have been matched by innovation in extraction, concentration and refining techniques. The technical problems of harvesting metal containing nodules, which are found spread over large areas of the seabed, are still too great for these resources to represent a serious threat to land-based deposits, but, as these problems are solved and as land-based extraction becomes more expensive, ocean floor extraction will become more competitive.

Undersea nodules containing manganese, nickel, copper and cobalt, as well as traces of other metals, are available (if that is the right word to describe material which may be several miles below the surface) in vast quantities. They cannot be mined or even dredged from the seabed but must be plucked or vacuum-suctioned from where they lie. As they are found mostly under international waters, they belong either to everyone or no one, but the harvesting of nodules has been referred to the Law of the Sea Conference.

The law of the sea

The Law of the Sea Conference (UNCLOS) sponsored by the United Nations was first convened in 1958 to deal with subjects like fishing rights and shipping conventions. Of the 150 countries participating in

UNCLOS 115 are from Third World countries but it is the Western nations, conscious of their growing dependence upon imports of metals, which have suggested that the nodules should be freely available to anyone with the capacity to extract them.

As the cost of developing the technology for extracting in this way will run into many billions of dollars, the Western world is effectively saying that only rich countries can benefit. Third World producers of these metals naturally disagree with this argument, for, not only will they not be allowed a share in the riches which are after all international property, but the metal gleaned from the seabed will replace the metal that they need to export in order to survive.

One possible solution to this problem has been proposed by the International Seabed Authority (ISA) which comprises all the nations of UNCLOS. They have suggested that a common fund should be established which would finance the commercial recovery of nodules for the benefit of all the participants.

Private companies, some of which have already invested large sums of money in undersea exploration and mining equipment, are suspicious of, if not hostile to, the proposal, not only because they feel that ISA will be given the choicest pickings from the seabed but that they may be obliged to hand over to ISA data and technology that they have accumulated.

At present there are only five major private companies with sufficient financial resources to be realistically hopeful of producing commercial quantities of metal from nodules by the mid-1990s. All five are consortia formed mainly from mining companies from USA, UK, West Germany, France, Netherlands, Belgium, Japan and Canada. They are all currently at the research and testing stage and are being frustrated in their efforts not only by the lack of agreement at the Law of the Sea Conference but also because they have no protection from each other. The massive fall in metal prices caused by the present recession may frustrate them further.

One other problem that they will have to come to terms with is that nodules consist of over 30 per cent manganese, a cheap and plentiful metal, with smaller quantities of other, more expensive metals. This means that if ever it becomes more profitable to produce, say, the world's cobalt from nodules rather than from land-based mines the market will be flooded with a vast surplus of manganese for which there would be no market.

The real criteria by which the world will judge the desirability of undersea mining will not be simple and will certainly not be based on cost-effectiveness alone. The Western world could easily be persuaded to encourage Western companies to go ahead with nodule exploitation even if the metals produced from them cost more than metals produced from conventional mines, as it would reduce dependence on what the

West thinks are unreliable sources.

If a significant proportion of the world's supply of certain metals were to be produced from nodules, it could mean that some poor countries would lose a major proportion of what little export earnings they have, thus making them even more politically and economically unstable. The West would, moreover, be paying a higher price than it needs to for these metals.

Speculation and Investment

The relationship between wealth and metals has long been established. Iron bars, gold, silver and copper have all been used as money, and metals generally have long been considered as a good hedge against inflation or a receptacle of wealth in times of major economic dislocation.

In modern times the vehicle for most investment into metals has been provided by the formal commodity markets specializing in metals trading. These markets were originally established to provide a link between producers and consumers and that is still, of course, their main function, but investors and speculators have long played an important part in commodity markets.

The net effect of speculative activity is said to reduce wild fluctuations in market prices. When the price of a metal falls to a low level because of lack of demand, the investor may feel that the price will rise in due course and will therefore buy the metal, thus preventing the price of the metal falling further. When prices rise to very high levels, investors may sell what stocks they have and even sell the metal short, thus preventing the price from rising higher. But this is not always the case in practice.

The effects of speculation and investment in metals are far from simple. In the first place it is by no means easy to distinguish the speculator or investor from any of the other people who use the market. The producer, for example, does not only use the market to sell the metal he has just produced. If he feels that the price of the metal may fall in the coming months he may sell forward his future production at a fixed price. Similarly, the consumer often buys metal in excess of his immediate requirements if he feels the price is likely to increase. Most metal brokers and traders are required to keep stocks by their customers and will adjust them according to their predictions of future prices.

In some exceptional circumstances, brokers and traders may find 'technical' ways of augmenting their profits by 'squeezing' or 'massaging' prices. If, for example a broker's research reveals that there is some bottleneck for physical supplies coming on to the market for a particular future date, he may take advantage of the situation by buying up a proportion of the metal available for that period. Other dealers and consumers may be forced to buy this material from the broker at a high premium over prices obtainable for metal available on a different date in order to satisfy a contractual obligation to the market.

Private investors and speculators who are not involved in the metal trade also operate on metals markets. Their presence in the market is generally welcomed by producers, consumers and brokers. Producers like to think that investors will purchase their metals when consumers are not buying. Consumers think that the investors' activity will prevent prices going too high too quickly and brokers like the commissions that investors have to pay the brokers for operating on the market. Their theoretical function then is to provide a kind of financial lubricant to the workings of the market.

There are times, however, when the activity of investors and speculators is far from welcomed. The most notorious case in the recent past concerned the machinations of the Hunt family from Texas in the silver market. Nelson Bunker Hunt had long been convinced that the price of silver was too low compared with the price of gold and sought to raise the price by buying vast quantities of his beloved metal. The price increase was so great that some producers held back supplies in anticipation of further increases. The dearth of metal reaching the market added new impetus to the spiralling price. By the beginning of 1980 the value of silver had increased by almost 2500 per cent since Bunker Hunt started his buying run. This was short-lived, however, as rumours spread that the Hunts had run out of money. The market collapsed. Accusations of market manipulation were met with counter-accusations. Brokers on the New York Comex market who had only been too happy to accept the commissions Hunt paid in his buying spree suddenly became frightened that purchases on this scale, however legitimate, would bring the market into disrepute. Hunt had failed to sustain the price of silver, and, ironically, it was the little old ladies, attracted by the high price of silver, who helped to collapse the price by flooding the market with their silver teapots and cutlery.

The Hunt episode is unusual in its magnitude, but the metal markets are often influenced by smaller, less well-publicized investment action. An individual or, more commonly, a consortium of individuals with sufficient funds and a conviction that a market is too low or high can and do manipulate prices through the exchanges although usually only temporarily but not always unsuccessfully.

Investment in strategic metals

At the beginning of 1980, some influential American investment advisers suggested that certain metals should be considered as a new investment medium. They coined the term 'strategic metals' to mean those metals which are essential in the manufacture of armaments and high-technology equipment, but whose supplies are vulnerable to disruption. Although these advisers had little knowledge of the metal markets, they did recognize that the USA was dependent on imports from unstable countries of many metals and that most of these metals were not traded on formal commodity exchanges.

The recent and bitter memories of OPEC's success in raising the oil price coupled with the 'resource war' theory, which was starting to be espoused at a high level at that time, was enough to create the momentum to turn the advisers' suggestions into action.

A number of companies were formed to provide investors with the facility to buy strategic metals. Several of these companies were, unfortunately, less than scrupulously honest. Metals like chromium, titanium, tantalum and germanium are only traded between producers, consumers and specialist merchant houses and the prices at which they are bought and sold are known only to people in the trade. It proved to be comparatively easy, therefore, to sell these and other metals to investors at two or three times their market price. Several of these companies have now been laid to rest by federal authorities and most of the business is now conducted by responsible organizations using some of the following arguments to persuade investors to consider strategic metals as a potential investment.

Production costs

The cost of producing metals is rising at a faster rate than inflation. Like all other industries mines are forced to pay more for labour, machinery, fuel, packaging and transport and higher rates of interest on loans. Moreover, because the deposits now being worked are less rich and mines more remote, producers have the added expense of treating more ore to recover less metal. The costs relating to infrastructure have also increased: roads and employee's accommodation are more expensive to construct and an increasingly high standard of service is required to attract labour to more inhospitable areas.

High technology

The advance of the electronics industry over the last decade has ensured a growing demand for obscure metals. Even a simple electrical device could contain a tiny quantity of a dozen rare metals for which there may have been no previous commercial use. Technological innovation in other areas such as catalysis, armament manufacture and atomic energy also require large supplies of rare metals with peculiar electronic, magnetic and physical properties. In spite of the recession in general industrial activity in recent times, demand for the rare metals for use in high technology will remain high.

Limits of supply

Supplies of metals are not infinite. Deposits of the rarer metals are spread more thinly over the world's surface than are those of base metals and many of them can only be recovered as by-products of the refining of base metal ores.

Seventy-five per cent of the world's supplies of columbium, used amongst other things for jet engine components, come from Brazil. Ninety per cent of the world's known reserves of tungsten are in China. Over half the world's cobalt comes from Zaire and Zambia. The problem of future scarcity of rare metals is thus compounded by the uneven distribution of future supplies.

Stockpiles

The United States Government holds a strategic stockpile of all the strategically important metals which is supposed to be equivalent to three years' annual US consumption. In practice stocks of some metals are greater and some smaller than this target. The new US administration has stated that these limits will be increased in some cases. A figure of $2.5 billion has been mentioned as the likely increase in expenditure on the stockpile over the next few years. Little is known about the stockpiles of metals held by the USSR but plans are underway to establish larger stockpiles of metals in the United Kingdom, France, Sweden and Japan.

These arguments are probably powerful enough to ensure that investment alone will be a major bullish factor in the markets of these metals. Private stockpiles are already being established and will be extended enormously once economic activity is revitalized.

The immediate future

The recession in industrial activity beginning in 1980 has proved to be deeper than any other slump since the last world war. The price of all metals has fallen as a result. As production costs continue to increase, profits from production of metals have fallen. Many mines now run at a loss. Production in Western countries is being curtailed. Mines and refineries close down every day and investment for future development is difficult to find.

Countries with centrally organized economies can afford to look at the situation slightly differently. The Soviet bloc as a whole is almost self-sufficient in metals and the fall in industrial activity has not been as steep there as in the West. Most socialist countries export surplus metals to earn much needed hard currencies and can be expected to export the extra metal they will not use because of the effects of the recession rather than reduce production.

Many Third World countries export very little in the way of manufactured items and must rely on the export of raw materials to earn enough to pay for the import of manufactured goods. These countries will be very reluctant to reduce exports of metal even if prices dip below the cost of production as the alternative is often to borrow money to which unacceptable conditions may be attached. These conditions often include reductions in public expenditure and the devaluation of the currency. Western companies which own mines in these areas are often put under great pressure to maintain high production at a loss.

The result of a prolonged recession may therefore be that mines in the West close, while production in the Third World and socialist countries is maintained or even increased, thus making industry in the West even more dependent on imported metals. Re-opening mines and refineries that have once been shut down is, moreover, a costly and time-consuming business, if such plants can be re-opened at all.

This may give the Third World a real chance of improving its relative wealth compared with Western countries when prices for its products revert to more reasonable levels after the recession.

Pricing of Non-Ferrous Metals

Twenty years ago the prices of almost all metals sold were controlled by the producers of those metals. There was no direct collusion on price strategy between the major producers, but producers' prices did tend to move in line with each other. What happened, say in the case of nickel, was that one leading producer would feel that the price of nickel could be increased by 10 cents a pound without reducing total sales of nickel. He would then inform all his customers of the change and post the new higher price. Within a week all the other major producers would follow suit. In periods of oversupply prices would be adjusted downwards in the same way. This method of changing prices ensured that competition between producers only occurred on the grounds of service, quality, reliability of delivery, etc, and not price. The prices of metals such as lithium, sodium and magnesium, the producers of which are few in number, are still very much controlled by the producers. But the control of producers over the pricing of most metals has broken down considerably.

Nationalism

Almost all the world's aluminium production was once controlled by only five companies. It has now become almost a status symbol for any country with an energy surplus (mainly OPEC countries but also countries with excess hydro-electricity) to build their own aluminium smelter.

Major mining companies have had to relinquish control of their operations in many countries. In many cases this has proved beneficial to the country concerned. It reduces the repatriation of profits out of the country and ensures that these profits are used for further developments or improved conditions for mining workers.

As production has moved out of the control of the giant multinational companies pricing strategy has moved into the hands of smaller producers. Small producers have little marketing experience and are less inclined to orchestrate their sales policy with competitors. This has meant that free market forces, the normal forces of supply and demand, have begun to dominate almost all world metal prices. The inclusion of aluminium and nickel in the range of metals now traded on the London Metal Exchange at free market prices would not have been possible even 10 years ago.

Producer price v free market

Producers who operate a producer price system enjoy some advantages but have many disadvantages compared with producers who simply sell at the best price they can obtain. In periods of shortage the free market price is at a premium to the producer price. This means that the producer price seller is assured of selling everything he produces but does not gain the advantage of the higher price. It also means that he must strictly allocate metal to his customers and threaten to cut off supplies to them if they should sell off their allocation at the higher free market price.

In periods of surplus the producer price is above the free market price and sales are difficult to place. The customer who received a good allocation of cheap material when there was a shortage often quickly forgets loyalty to his supplier when the customer's competitors have the advantage of buying cheap free market material. The producer is forced to reduce production or stockpile his unsold metal. If the free market price continues to fall the producer price must be reduced to placate customers. An announcement of such a move usually helps to reduce the free market price still further.

The arguments used by the producer to justify the producer price system is that such a policy ensures a price to both producer and consumer which is reasonable and will be maintained at the same level for some considerable time. This assists both the producer, who may wish to plan new investment, and the consumer, who can quote fixed prices for his product for future delivery. These advantages can, however, be obtained on a constantly moving market by using hedging techniques. Producer prices are just as unpredictable as free market movements and, whereas the free market moves constantly but by small increments, the producer price usually moves instantly and by huge leaps.

The future of metal pricing systems

Producers' cartels have been singularly unsuccessful in metals markets but a prolonged period of recession may change the situation. Producers have less incentive to work together when metal prices are high and, whereas cartels between Western-based multinational companies are difficult to justify, cartels between producers in developing countries could be a way of improving the balance of wealth between rich and poor nations.

The success of such cartels will depend to a large extent on the metal concerned. If most of a certain metal is produced in Third World countries with a similar political ideology a cartel in that metal may be successful.

There are two possible ways in which this could be achieved. CIPEC (the copper exporters' organization) has always attempted to raise prices by trying to reach an agreement between the members on reduced production. Such agreements have never been detailed enough to ensure success. Does a cutback in production mean a cutback in mined ore or finished product? In either case is it a cutback from the previous year's production or 12 times the average of the last month, and how are existing stocks to be taken into account? Such agreements are more complicated than they seem and need a great deal of detailed planning and rules which must be strictly abided by in order to assure success.

Efforts to raise prices by cutting production must inevitably be accompanied by the laying off of mining workers, thus adding to the burden of poor countries where unemployment is usually already at an unacceptably high level. So, work sharing or compensation agreements, which can be paid for out of the higher revenue from the sale of the metal, must be incorporated into the plan.

Another method of increasing prices for a metal, often favoured by miners, incorporates the use of a 'buffer stock'. By this method, minimum and maximum price levels are agreed to and set by the producers. If the price falls below the minimum price, the producers build up a stockpile of surplus material and purchase metal from the market until the minimum price is reached. If the price moves up beyond the agreed maximum level, the stocks are sold until the price falls back below this maximum limit. This method is currently used to control the tin price under the International Tin Agreement to which both producing and consuming countries are party. A similar method has been used by Alufinance to raise the price of aluminium in Europe.

The major disadvantage to producers using the 'buffer stock' method is the difficulty of finding the initial funds to purchase and stockpile metals in the period of oversupply and low prices. Furthermore, there is nothing to prevent the price rising above the maximum agreed limit if the 'buffer stock' is exhausted. This, of course, only represents a problem to producers if they wish to have consumers participating in the plan.

In theory, there is nothing to prevent either of these methods being used by metal producers, but, in practice, many obstacles need to be overcome. Firstly, many poor countries are completely dependent on consuming countries for aid and loans, the terms of which could easily be altered to discourage a country from joining a cartel arrangement. The greatest obstacle so far experienced by such cartels seems to have been overcoming differences between members of the cartel and engendering an atmosphere of mutual trust. A long period of low prices might be all that is needed to convince producers that these problems must be overcome.

Substitution

The market determines that no metal can sell for more than its effective substitute. As the price of a metal increases, more efforts are made to find substitutes, but often substitutes cannot be found or the substitutes can only be used for a small range of uses and may not do the same job as effectively.

The relationship between the price of a metal and its substitute determines the ratio of those two materials used for the same purpose. An interesting case in point is the relationship between magnesium and aluminium. Both metals have very similar physical properties and have a large range of uses. Although both metals are light, a unit volume of magnesium is only two-thirds as heavy as the same unit volume of aluminium. For this reason both metals are used wherever lightness is required, ie in components for aircraft and in diecastings for motor cars where weight saving means fuel saving.

Magnesium is about twice as expensive as aluminium but, if magnesium was only one and a half times as expensive, wholesale substitution would take place in favour of the lighter metal. Because this price ratio has never yet occurred magnesium production has been kept low. In fact only one ton of magnesium is produced annually for every 20 tons of aluminium. Magnesium is at present only used when weight saving is absolutely essential (or in certain other minor uses), but, if more of it could be produced, the unit cost of production would fall. It is quite probable that in 20 years, when fuel saving will be even more important, magnesium will be a more familiar metal than aluminium, especially as the source of magnesium (seawater) is extremely abundant whereas bauxite (aluminium ore) is becoming increasingly expensive.

This is only one illustration of the relationship between substitutes. Non-metallic substitutes for metals are also common. Plastic and glass containers have replaced some tin and aluminium cans. Certain ceramic substances can be made into magnets. Organic chemicals are used to replace selenium and cadmium as pigments, and carbon fibre has replaced some high-strength, low-density alloys used in jet engines.

The substitution of one metal for another is, however, more usual. Substitution is not always made on the basis of cost. Substitutes for mercury and cadmium have been found in many industrial areas because both metals are extremely toxic. Spent uranium has largely replaced bismuth in the manufacture of acrylonitryl simply because there was not enough bismuth available.

Changes in technology also affect the use of, and substitution between, metals. Substituting the transistor for the electric valve has reduced the consumption of some metals and increased the use of others. Optic fibres will replace copper wire but increase the use of germanium, a metal used in the production of optic fibres.

How Metals are Traded

Most metals are sold directly from the producer to the consumer and there are a variety of ways in which this is done. For the direct sale of a metal traded on one of the world's official metal markets the value is usually based on the price traded on the exchange at or about the time of physical delivery. This not only applies to the metal itself but also to the sale of the ore, concentrate or alloy of that metal. A typical sales contract would specify the total tonnage to be delivered over a given period and in how many shipments. It would also give a price formula which would describe the relationship between the price to be paid and the price of the metal on the exchange where it is traded.

Where the material is to be delivered in the form of ore, concentrate or scrap the price would naturally only be a proportion of the traded price. Where the metal is to be delivered in a high-purity form, semi-fabricated form or alloy, or where favourable payment terms apply, the price to be paid would be at a premium to that ruling on the market at the time of delivery.

Framed contracts

Where large quantities of any metal are sold, sales arrangements are usually made under a 'framed contract'. Here most of the terms of sale are specified, ie total tonnage to be delivered in the year, maximum and minimum tonnage to be delivered each month and in whose option, port of delivery, packing, etc. The only item missing from a framed contract is the price.

The contract only becomes valid when the seller and the buyer agree, by negotiation, on a price for deliveries over the next three months or whatever period the contract specifies. Although such contracts are common, especially for bulk cargoes like bauxite, manganese ore, chromite and iron ore (ie metals with no official market), both parties need enough goodwill to negotiate a reasonable price.

Agents

Small producers are often unable to afford a marketing department which is capable of negotiating with all their customers, in which case agents are employed with specialist knowledge of both the product and the market in which it is to be sold. Some companies act only as agents for producers, but most agents are themselves merchants or consumers of the materials concerned.

Agents usually work on a commission which represents a percentage of the value of the material sold. Such commissions vary with the amount of effort and risk that the agent is required to undertake. Where an agent is given the task by the producer of selling a new product or selling into a new market, it is normal practice for the agent to be paid a 'retainer' or lump sum to cover the first year of the agent's expenses. This sum is then usually deducted from the total commissions earned by the agent in that year.

Metal merchants

The producer often prefers to have no direct link with the ultimate consumer of his product and may even wish to avoid a link through an agent. In such cases the producer relies upon metal merchants to purchase his product. The world's largest metal merchants are based in traditional centres of commercial activity – London, New York, Switzerland, Dusseldorf, etc – and can often offer the producer better terms than could be achieved either by selling directly to consumers or through agents.

The merchant may, for example, feel that the price of a product is likely to increase, in which case he will buy the material from the producer without immediately bothering to make corresponding sales. He knows that by 'taking out' this material from the market for some period the chances of prices increasing will be further improved.

The merchant also eliminates the risk to the producer of a customer's possible financial failure. A large merchant may also be able to offer advanced payment for goods which the producer could not possibly expect to get from a consumer. Although most metal merchants neither produce nor consume the metals in which they deal, they have the advantage of being both buyers and sellers of the product, which gives them a much greater insight into market activity and trends.

Consumers rarely talk to other consumers and producers rarely talk to other producers but merchants talk to both. In fact, a merchant's job consists largely of constant and global communication not only with the ultimate buyers and sellers but also with other merchants with whom they trade. This enables the merchant to buy at the cheapest levels and sell to buyers in the greatest need.

Tolls and swaps

A growing aspect of merchant activity is the toll conversion of base products into refined or semi-fabricated products. At any given time some mines produce more material than can be conveniently refined by them and some refineries are not able to find enough raw material to meet their refining capacity. The merchant's thorough knowledge of refining activity often enables him to toll convert (that is, to pay for the conversion of the material) on profitable terms.

A producer is often only able to supply metal at the wrong place or at the wrong time for one of his customers. A merchant may know of a second producer who has the same problem but whose metal would fit the requirements of the first producer's customer. If the first producer's metal can be sold easily by the second producer, the merchant can make a simple swap and be paid a commission for doing so.

It may sound, from this simple description, that the chances of finding such a neat solution to the problem are rather remote, but, when one considers the vast number of suppliers for any given metal, the odds against finding such an arrangement fall sharply, so much so that several merchants make their living by doing nothing else.

The broker

On the world's two most active metal markets, Comex in New York and the London Metal Exchange in the UK, brokers act for producers, consumers and investors. Many of the buying and selling orders they receive in a day can be matched against each other, eg a producer wishing to sell 100 tons of copper at the three months' price will be matched off with a consumer who wishes to buy 100 tons of copper for the same three months' delivery time. The price of the metal traded during the period in which both deals are to be executed may change. It is therefore up to the broker to decide which price within the range is fair to both buyer and seller. The broker will receive his commission for both deals.

Brokers on both exchanges compete with each other and try to establish close relationships with clients, but most clients use two or more brokers often at the same time.

Commission rates are not fixed and brokers will often be persuaded to reduce rates in order to obtain a large client. It is, however, not always wise for a client to choose the broker with the lowest commission. If a broker has several selling orders for a particular metal to be executed in the same time limit and the price of the metal moves during that time, the broker has the choice of matching up a particular client's order with a particular execution on the exchange. An important client paying high commissions is more likely to find that he has sold at a higher price than the client paying only a small commission. Some clients therefore take the precaution of connecting themselves by telephone with the broker's clerk who is standing on the floor of the exchange reporting price changes as they occur. This will at least reveal whether a broker is doing a good job for him. If he is not he can change his broker.

The international market

Considering the enormous economic importance of metals, the number of people actually engaged in international metal trading is relatively small, numbering perhaps a few thousand. But it is by the prices and conditions of purchase and sale of metals arrived at by these people at any given time that the world market is established. Although the forces of supply and demand, glut and shortage dominate the way in which prices move, sympathy and psychology also have an important part to play.

Most of the world's important dealers in a particular metal know each other well. They talk to each other regularly. They meet on the exchange and at conventions. They visit each other and exchange views. It is therefore not surprising that a kind of collective opinion about the prospects for a market is quickly established. A quick glance at the price graph of any metal will show that this opinion is not always correct. Small peaks in the graph often appear in the middle of a downward trend. These peaks are often caused by sympathy alone. If, for example, a group of merchants agree that prices should be higher it is not long before they have influenced their friends into thinking the same way. Eventually consumers and investors hear about it. People with short positions buy metal in case the opinion should prove to be correct. This causes the market to rise. Consumers raise their stock levels. Investors take a punt. The opinion turns into a self-fulfilling prophesy. The price rises. But if stocks of unsold metals are larger than had been estimated or producers quickly increase production to gain the advantage of the higher price, the price upsurge is short-lived. Investors get worried and sell and the market collapses. So the price has risen and fallen without any fundamental change in the market.

Such occurrences are common and show that the price of metals can sometimes reverse the principles of supply and demand. People tend to buy more not less when prices rise and vice versa. These short- and medium-term trends are, however, stabilized in the longer term, where stronger market forces always dominate.

Summary of the Non-Ferrous Metals Market

The base metals: copper, zinc, lead, aluminium

The markets of these metals have similar characteristics in that they all have a very wide range of uses and a wide distribution of sources. Their prices reflect the world's industrial activity and there have been no major disruptions of supplies for some years. Since the beginning of the decade the world's economies have been put under great strain by high interest and inflation rates and wildly fluctuating exchange rates which have resulted in a reduction of activity in those sectors of industry which consume these base metals. These sectors include automobiles, machine tools, shipbuilding, construction, mining and domestic appliances. Demand has always been a more important factor in determining metal prices than the cost of production. Some base metal producers would prefer to continue production of metals which can only be sold at a loss rather than liquidate massive investments, but investment in future production has been cut drastically.

It is still too early to predict how long the present recession will last. Certainly there are as yet no signs that the efforts being made to cure the underlying problems of inflation and low growth are working. But if new moves are made to stimulate output the future capacity of base metal production is so limited that we should expect price rises which will make previous booms in the prices of base metals look insignificant. Fresh investment in new base metal production capacity will only be forthcoming if investors can be assured of sustained growth in consumption at prices which will guarantee a return on costs of developing new mines and plant which could be two or three times the level of old existing plant. These high prices may themselves be a constraint on any industrial recovery in the sectors that consume them.

The refractory metals: molybdenum, tungsten, vanadium, columbium

The major use of all these metals is in the manufacture of special steel alloys. Prices of all these metals have increased dramatically over the last few years as innovations in technology have demanded that engineering components must stand up to greater speeds, pressures, strains and temperatures. China's control over the tungsten market enabled it, acting almost unilaterally, to increase tungsten prices in the late 1970s. A shortage of molybdenum supplies occurred soon after when large quantities of the metal were needed to make steel alloy oil pipelines as newer, more remote oil fields were opened up, stimulated by the huge increase in the price of oil in the early 1970s. Vanadium and columbium are physically very similar to molybdenum and can be substituted for it in many uses. As this happened the prices of these two metals rose as well. Plans to increase molybdenum production were accelerated as prices rose but increased production coincided with a fall in demand caused by the recession and molybdenum prices collapsed. It is now assumed that supplies of all these metals are sufficient to meet even a major increase in consumption over the next few years and prices will remain stable. China is, however, still in a strong position to increase tungsten prices if demand increases.

The steel alloy metals: chromium, nickel, manganese

The less exotic steel alloys made with these metals are, like base metals, used in a large range of industrial applications. For this reason consumption has fallen during the present recession. Approximately half the world's supply of chromium and manganese comes from South Africa and the USSR. This has led to concern that supplies could be restricted for political reasons. In spite of this possibility the price of both metals has remained so low that production has had to be cut and plant for turning the ores of the metals into ferro-alloys has been closed down on a massive scale, especially in Europe.

The price of nickel could have fallen to lower levels if there had been no threat of a strike at INCO's massive Canadian plants in 1982. There is no immediate likelihood of increased consumption of these metals, and no major price increases can be expected for some years.

The light metals: lithium, magnesium and titanium

Most lithium is used in the manufacture of certain chemicals and the salts of magnesium and titanium have important uses, but the metallurgical application of all these metals is in areas where weight saving is needed. These are not only in the aircraft industry and for making missiles but also and increasingly in automobile production. Civil aircraft production has slowed down since the very busy period of the late 1970s but expenditure on military aircraft has remained high. Many applications for these metals are being found to reduce the fuel consumption of vehicles by reducing weight. The lithium and magnesium markets are almost totally controlled by the major producers and prices can be expected to increase in line with inflation, but so much extra titanium production capacity has been put onstream since the recent titanium shortage that prices are not likely to increase in real terms until well after the end of the present recession.

The minor metals: cadmium, bismuth, selenium

In the late 1960s and early 1970s all three of these metals were in short supply. But they can now be bought at a fraction of the price they reached in this period and are cheaper than ever before. All three are produced commercially as by-products in the refining process of base metals. At one time the refiners of lead, zinc and copper looked upon these metals as a lucrative bonus in their activity, but as prices have slumped they are now considered as contaminants which must be sold at below the cost of extracting them. It is likely, however, that new uses will be found for these metals now that they are so cheap and prices cannot be expected to remain so low for long.

The electronic metals: germanium, indium, gallium

The major uses of these three rare and expensive metals are in the electronics industry where they have a host of applications. Prices have dipped a little from the peaks they reached recently but demand continues strong. Electronics is one section of world industry which has suffered comparatively lightly in the recession. The increasing use of the microprocessor, laser technology and fibreoptics in the home, in transport, and in factories, shops and warehouses will sustain the growth of the electronics industry.

Electronic systems require components made with these and other rare metals but changes occur so quickly in the electronics industry that the use of a particular metal in some applications can be abandoned completely in favour of another in a matter of months. Large quantities of germanium, for example, were used in the production of light emitting diodes, but these were phased out in favour of liquid crystal display systems and the consumption and price of germanium fell. But a new and significant use was found for germanium in the manufacture of infra-red nightsighting equipment, and this has produced a shortage of the metal. Recently another new use for germanium in the production of optic fibres has been found, which seems certain to exacerbate the critical shortage.

Other metals

In spite of competition from other materials such as glass and aluminium used for storage and packing of food and drink, tin which is used mainly for tin cans has remained a buoyant market. There seems to be little doubt that the main producing countries have achieved this by buying surplus material from the market outside the terms of the International Tin Agreement. This must have been a very expensive exercise but has been rewarded with a price for the metal which must be satisfactory to the producers. The battle between the producers and consumers for the tin market seems likely to be fought outside the ITA with the USA using its stocks to relieve shortage and the producers taking over more control of the ITA.

Cobalt producers have been unable to coordinate their efforts to take advantage of the high prices which consumers had been accustomed to paying in 1978 and 1979. Stocks in producers' hands have built up to such high levels that it can be safely assumed that only a massive increase in world industrial activity would justify a price rise.

The production of arsenic and mercury has been so circumscribed by low prices and the problems of toxicity that cutbacks are inevitable. The weight of evidence that certain heavy metals are poisoning the environment is now so overwhelming that new moves to restrict both production and consumption of these metals will continue, but in the meantime production has fallen faster than consumption creating a slight shortage of these metals. Production of both can be quickly increased, however, if the price was high enough to justify the expense of building production equipment which would overcome any environmental hazard.

The Role of Non-Ferrous Metals

Of the 100 or so known elements over 80 are metallic. Each metal has its own definite physical characteristics but not all metals have found an important role in industry. This may be because no use has yet been found for them, or that another cheaper metal with similar properties is employed, or that the metal is so poisonous that it cannot be mined, processed or handled without creating environmental hazards.

Conversely, there are some metals, even quite rare ones, that have achieved a massive importance in technology and are indispensable in the modern world.

Part 2 covers those metals that are commercially important enough to sustain a regularly traded market but it excludes the precious metals, that is gold, silver and the platinum group metals. The markets of these metals are governed as much by currency fluctuations and investment considerations as by supply from mines and consumption by industrial consumers.*

Each metal dealt with in this book has its own market idiosyncrasies. For anyone who wishes to follow these markets, this book can only be a primer and a guide. In order to understand why a market behaves as it does on a given day the various metal trade periodicals need to be studied and there should be regular exchanges with merchants who daily trade the particular metal.

Price influences

Prices are affected by demand, supply, exchange rate fluctuations, strikes, wars, technical innovations, environmental considerations, the price of fuel, changes in stockpile limits, new mines, freight rates, and, perhaps more important than all these in the short term, the judgement or even the whim of the merchant, producer, consumer or investor.

The smaller the market for a metal, the more susceptible it is to manipulation. Selenium, for instance, has a world production of only about 1,000 metric tons. At $5.00 per lb it would theoretically cost less

* See Robbins, Peter and Lee, Douglass (1979) *Guide to Precious Metals and their Markets* Kogan Page Ltd: London

than $12 million to buy the entire world output of this metal with no chance of production being increased because selenium is a by-product of copper production. Selenium is essential in the production of photocopying machines. Could it be that for the comparatively small investment of $12 million, an investor could threaten to bring to a halt the world's photocopying machine industry, or at least force that industry to pay almost any price for its selenium supplies?

In practice, this could not quite happen because of the difficulty of buying enough metal without others knowing about it. But by buying 20 or so tons of selenium at a time, the market would go up by at least $1.00 per lb. This type of risky speculation has taken place many times, particularly in periods of natural shortage.

Most of the metals dealt with in this book, however, are not by-products. They are mined in their own right. But for many of these metals this has not freed the markets from short-term manipulation by merchants and producers alike.

Surprisingly, perhaps, there is little investment interest in metals that are not traded formally on one or other of the world's commodity exchanges. This may be due partly to the volatility of the markets or to the specialized knowledge required to deal successfully.

In the copper market, for example, non-metal trade investors, including banks, insurance companies and pension funds, are prepared to support the market when its value goes too far below cost price. This is because copper is traded on a futures market which has had a good record of utilizing available investment funds.

There is, however, virtually no chance of including many rare metals in such formal commodity exchanges in the foreseeable future. The worldwide turnover is far too small to justify the cost of such an operation.

The markets of some of the rarer metals are greatly distorted in relation to the true supply/demand factors. These metals are by-products of the production of other less rare metals, which means that however great the demand for the by-product, production cannot be increased to meet this demand without a similar rise in demand for the major product.

There is an apparent limit to this state of affairs when the value of the by-product exceeds the value of the major product in a given volume of ore, but this is an extremely rare occurrence as most by-product metals only constitute a small percentage of the major production.

The application of these metals is often extremely sophisticated. They may be part of an alloy forming a minute but integral part of a jet engine, or chemical process, or an electronic device in which the value of the metal used represents an insignificant part of the cost of the entire system in which it is used. This means that if no substitute is available, the price of the metal can rise by several hundred per cent before it has any effect on the competitiveness of the consumers' product. These factors help to create erratic, volatile markets.

In times of high demand, producers are unable to increase output and consumers are not concerned with the price they pay. In the recent past, the price of cobalt, cadmium, molybdenum and indium (to give typical examples of by-product metals) have displayed price rises of around 500 per cent in times of high demand which have lasted a few months at a time. These metals have markets that are more volatile than any other commodity or investment market in the world.

To add to the difficulties of predicting future consumption and the uses of such metals, there are other factors to consider. For example, technicians are inventing new uses for rare metals every day, particularly in the electronics and chemical industries. An obscure piece of research could, one day, result in an important new use for a rare metal as a semi-conductor or as a catalyst. Furthermore, reserves of metal in the earth's crust are finite and the costs of extracting, refining and distributing metals rise each year, with a consequent increase in the risk of investment in metal production.

Alloys

Very few metals are used in their pure state. They are used instead as alloys: a combination of two or more metals. There is an almost limitless permutation of alloys but the most important are those consumed in massive quantities by the steel industry. These are known as bulk ferro-alloys and are the alloys of iron with chrome, silicon and manganese. The markets for each of these alloys are so large that they exceed the markets for most pure metals and represent the most commonly traded form of the three metals concerned. Other ferro-alloys, such as ferro-nickel, ferro-titanium and ferro-molybdenum, are also important enough to have their own independent market.

The only other types of alloy to have an international market are those based on zinc, copper, nickel and aluminium.

Part 2:
The Metals

Aluminium

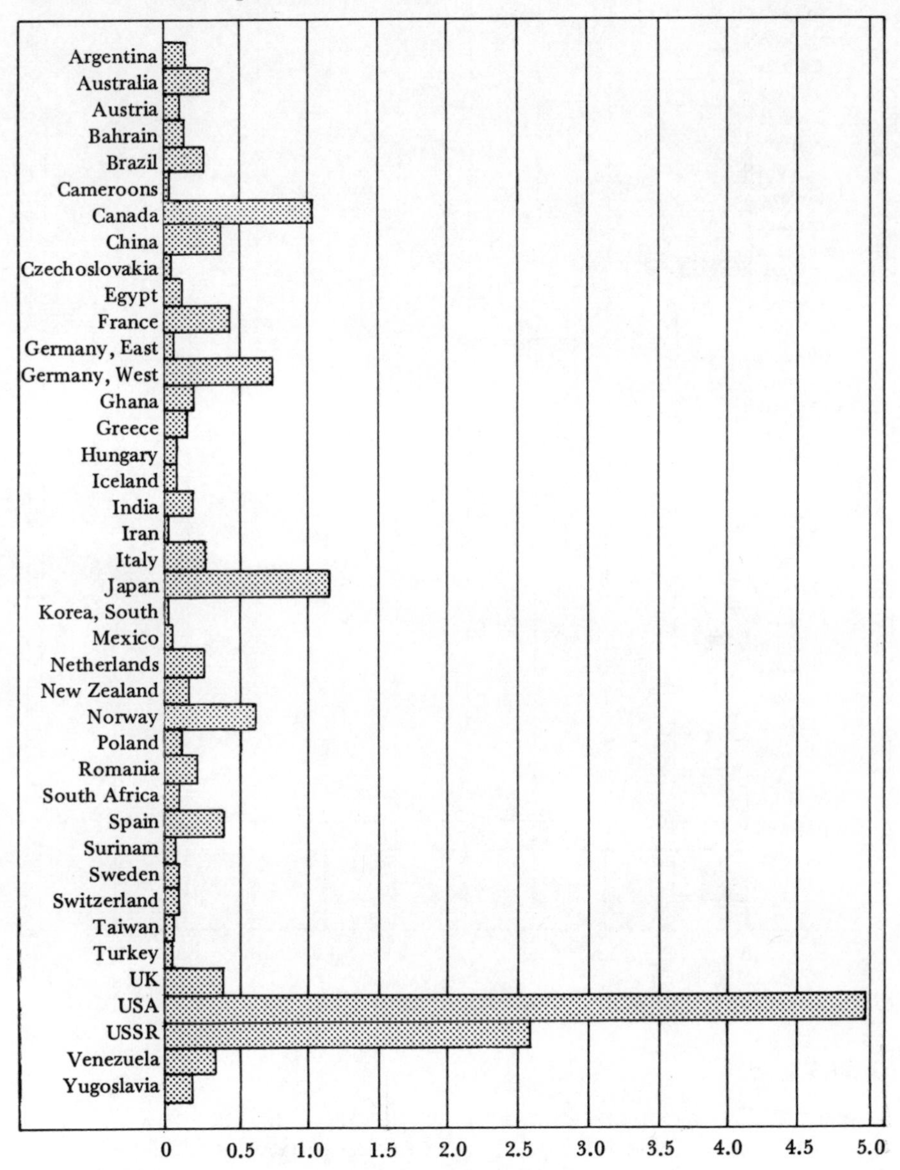

Total world production 16,496,600 metric tons

Aluminium metal consumption, 1980 *million metric tons*

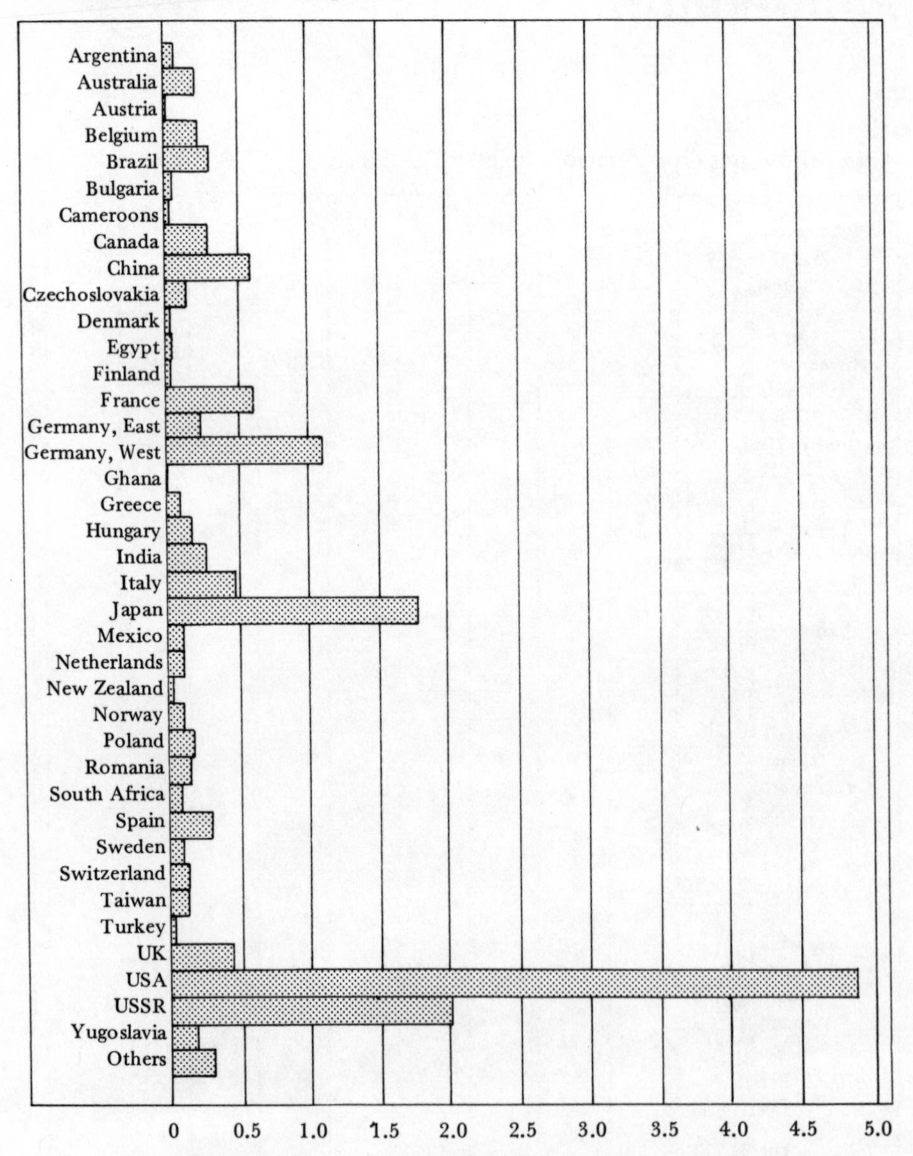

Total world consumption 15,494,700 metric tons

Grades available

The bulk of unwrought aluminium is produced in the form of ingots weighing approximately 22 kilos with the purity of 99.5 per cent minimum or 99.7 per cent minimum. Many smelters, however, can offer a large range of shapes and grades, including 'T' bars – up to 1,000 kilos each; sows, 200-500 kilos each; extrusion billets; rolling slabs and wire rods.

Most modern smelters are designed to produce metal with the minimum purity of 99.7 per cent with iron content below 0.2 per cent and the silicon content below 0.1 per cent.

Production method

Aluminium is the most abundant metallic element in the earth's crust. The metal is extracted from alumina (aluminium oxide) which is produced from bauxite, a naturally occurring ore.

It is produced by the electrolysis of alumina with cryolite, a natural mineral that can also be made synthetically. The anode of the electrolytic production cell is made of carbon which combines with the oxygen in the alumina when the current is passed through and sufficient heat is generated, reducing the alumina to the metal. The molten metal is then cast into the moulds required.

Major uses

More aluminium is produced than any other non-ferrous metal, and none has such a diverse range of uses. The metal has a high thermal and electrical conductivity and a high strength, particularly when alloyed with small quantities of other metals such as silicon and manganese. It is extremely ductile and has found many uses as sheet and foil, and can be cast comparatively easily into all kinds of complex shapes.

Its resistance to corrosion and its attractive appearance makes this metal suitable for many applications in the construction industry. Its lightness (aluminium is less than half as heavy as steel) makes it the most important metal in the aircraft industry. The versatility of aluminium explains its phenomenal growth rate.

It competes with copper in its application as a conductor of electricity, with steel in construction, with zinc for castings and with stainless steel in its corrosion free properties. It is also used for 'deoxidizing', that is the property of removing oxygen from molten iron and steel. There are few manufacturing and industrial activities which do not have a substantial use for this most useful of metals.

Main market features

Because of the economic and political instability of some of the main bauxite producing countries, many aluminium manufacturers are investigating the possibility of producing the metal from other minerals containing aluminium, of which there are many. These include china clay and shales, and good progress is being made in this field. These efforts, however, have been interpreted as a deterrent to any potential threat from major bauxite exporters who might wish to form a cartel to force price increases.

It has become almost fashionable of late for an emerging nation to attach great priority to the building of its own aluminium smelters. This is particularly true of countries in the Middle East, South America and South East Asia. The pattern usually includes the majority shareholding held by the country concerned with a minority interest held by a large aluminium producing company which provides all the technical expertise to construct and run the smelter.

The recent history of the industry has been marked by the failure of the market to predict future consumption trends. This has resulted in a large degree of over-capacity in times of stagnant demand, and shortages in times of high demand. The phenomenom is partly due to the long period (about five years) from the date of conception of a new production plant, to the date of commencement of production.

Predictions of the aluminium consumption growth rate vary from 4-8 per cent per annum over the next decade, a very high growth rate for a metal, whichever forecast is used. It should be remembered that, although the price per metric ton of aluminium may seem close to the price of any one of its competitors, the lightness of the metal means that one metric ton will go much further: it has greater volume than any of its competitors, with the exception of magnesium.

Aluminium is essential in the manufacture of highly sophisticated items such as computers and aircraft, but is also used for making commonplace items such as cooking pots. Its consumption, therefore, is not much affected by technical changes in any industry. The market for aluminium is affected by fundamental factors including the supply of raw material, general world economic activity and competition from other metals.

By and large production is expected to keep pace with consumption but a high factor (about 10-15 per cent) of the cost of production is energy. The ever increasing cost of construction of smelters and of raw materials will inevitably result in an average rise in price. Fluctuations in price from year to year may be more extreme as control of the market shifts from the very few large producers to the many new ones.

Known world reserves (bauxite)

5,842,000,000 metric tons. This is equivalent to 1,168,400,000 metric tons of aluminium metal.

Method of marketing and pricing

The major producers publish an official price for aluminium, the most important of which is the American price. All American producers, at least theoretically, abide by this price. The price is for virgin, unwrought ingots or sows with a minimum purity of 99.5 per cent, CIF major Western ports. There are premiums for a higher purity, different shapes, delivery to works and other ports, or for the delivery of molten metal in specially designed lorries. Most major manufacturers are vertically integrated, producing everything from ingots to foil to manufactured items, often through subsidiaries and associated companies. They are frequently their own largest customers.

Similar official pricing systems exist in the UK, Japan, Continental Europe and other industrialized countries, but in times of glut, producers are often forced to sell at discounts on their official price. These discounts are often hidden in extended payment terms or other similar devices.

There is a growing free market in aluminium, and in October 1978 the London Metal Exchange introduced an aluminium futures contract that has established itself as a pricing and hedging medium despite initial strong opposition from leading producers. However, the Metal Exchange market is not significant for industry as UK producers retain control over prices and marketing. The free market is conducted by international trading houses, which generally have no financial connection with producers. These merchants, however, may have a close working relationship with major smelters and could, for instance, conduct 'location swops' or 'time swops' between producers. That is to say they find one smelter with the stock at point A and the sale at point B and another smelter with the stock at point B and the sale at point A. They simply arrange a swop, thus saving transport costs, and also collect a commission. If you assume A and B to be times, such as March and July, the swop on this basis is a 'time swop', which saves one company's interest costs. The merchant may also arrange tolling contracts, that is to say he may arrange with the smelter to utilize surplus production capacity to convert the merchant's alumina into ingots for a fixed price per metric ton.

The bulk of free market activity is concerned with the trading of metal from small producers who have no fixed pricing policy and insufficient international marketing arrangements. They must sell to merchants at the best possible price, but in times of shortage they have the advantage in that they are able to obtain higher sales prices than the major producers who abide by official price policies.

This merchant activity accounts for about 10 per cent of the world's aluminium trade and the prices recorded for these transactions give the best guide to market trends. Aluminium merchants are active in most major commercial centres of the world.

Scrap and secondary aluminium

Many of the useful aluminium alloys can be made from scrap simply by melting it down, adding other metallic alloying agents and perhaps a percentage of pure aluminium, to produce the right specification. This has ensured a very active scrap market. Indeed, the scrap price may be at times as much as 85 per cent of the pure aluminium price.

World secondary aluminium production is approximately one third of primary production.

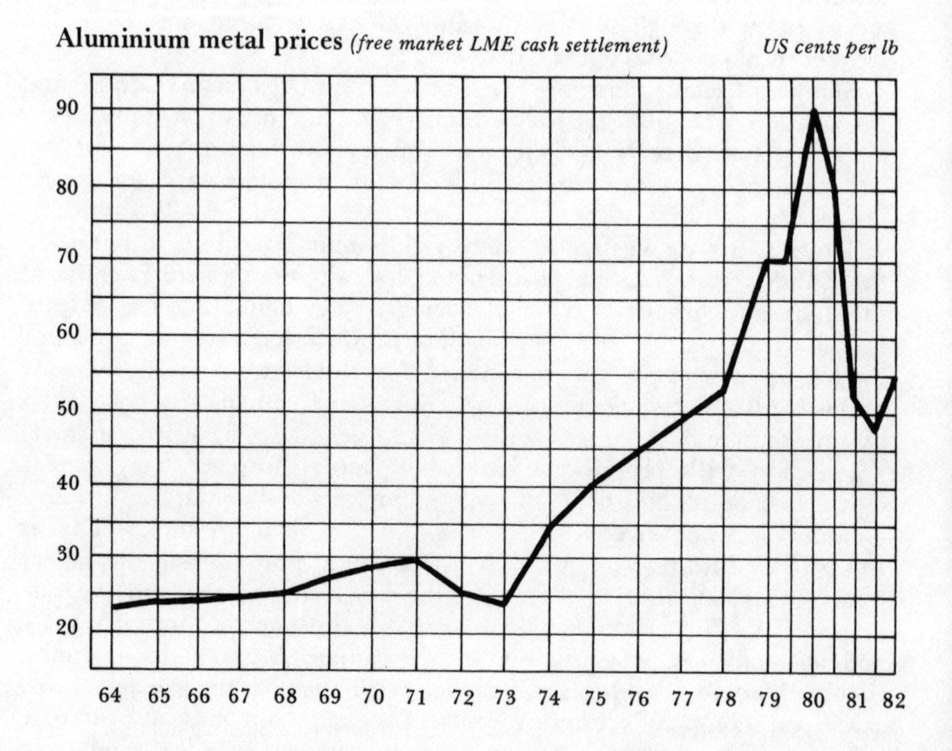

Antimony

Antimony metal production, 1979 *thousand metric tons*

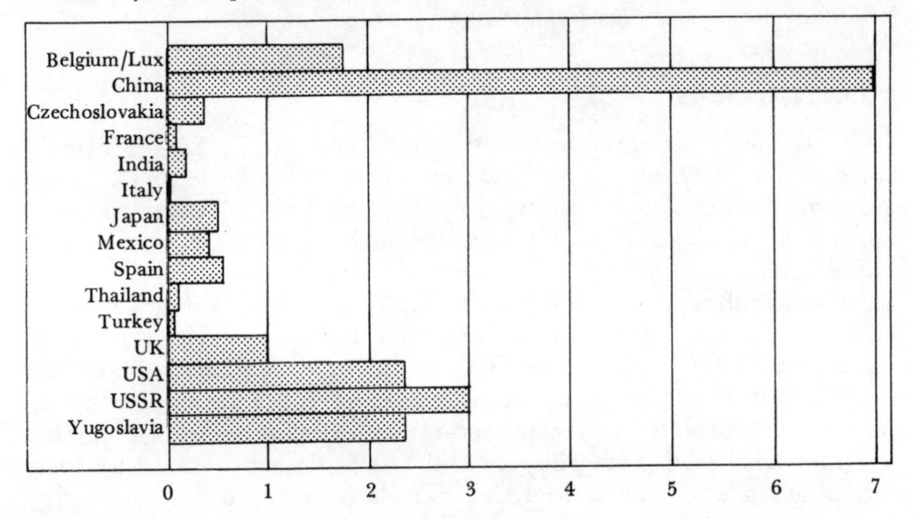

Total world production 20,223 metric tons

Grades available

Antimony metal is normally marketed with the purity between 99 per cent and 99.85 per cent, the most common grades being 99.5 per cent and 99.65 per cent. Particular attention is paid to the arsenic content which most consumers require to be below 0.2 per cent. The normal Chinese grade of 99.65 per cent minimum and 0.15 per cent As maximum has become the most commonly traded grade. The metal is traded in the form of ingots which weigh between 20 and 50 pounds.

Production method

Antimony is produced from its ore, stibnite – antimony sulphide – by roasting to oxide, followed by reduction to the metal by heating with coke in a simple process. Antimony concentrates, derived as a by-product from the production of other metals, mainly lead, are usually in the sulphide form and are dealt with in the same way.

Major uses

By far the largest use for antimony, half its total consumption in fact, is in batteries where it is alloyed with lead. Its effect on lead is to make it more robust and to reduce corrosion by chemicals. The metal is also combined with lead and other metals in the manufacture of solders, print metal, electric cable sheaths, and in certain types of ammunition. Its oxide is particularly important in the manufacture of flame retardant paints and plastics, and as white pigment.

Main market features

China is the largest producer of the metal and has, by far, the largest reserves. Its dominance of the market is having a major effect on large Western producers, particularly in the US and the UK. These producers appear to have decided to concentrate on the production of antimony oxide and the recycling of antimonial lead, mainly from scrap car batteries. Although they still produce a little metal for their domestic customers, they leave this market mainly to the Chinese. Antimony oxide is made directly from the ore, but there have been a few temporary occasions in recent times, when market circumstances have allowed producers to manufacture oxide profitably from the metal.

Antimonial lead is an alloy of between 3 per cent and 10 per cent antimony in lead and is used to make the plates in batteries. A large proportion of this material comes from treatment of used and scrap batteries.

Consumption of antimony by the automobile industry is a dominant feature of this market, not only because of its use in car batteries, but also for its use in flame retardant plastics needed in cars.

There are substitutes for antimony in all its major uses. Other inorganic chemicals are used as flame retardants and these are cheaper and less toxic. However, antimony is still used as a flame retardant in the plastic insulation of electric cables as the bulk required of the alternative flame retardant affects the plastic's mechanical qualities.

Calcium combined with a little tin can and does replace antimony in car batteries and some major car manufacturers are incorporating this feature in new models, in spite of the fact that calcium-lead batteries are more difficult to charge.

Known world reserves

Approximately 3,550,000 metric tons of which 2,000,000 metric tons are estimated to be in China.

Method of marketing and pricing

There is an active free market in antimony metal ore and trioxide but the prices of these three items have little relationship to each other in the short term. The antimony metal market is rather volatile and attracts speculation, mainly from the merchants who deal in it. This feature is reinforced somewhat by the unpredictability of the Chinese marketing policy. Virtually all Chinese production is sold to international merchants who, in turn, distribute to consumers (many of the major consumers in the free market are in Eastern Europe).

Japan, a major consumer of antimony in the form of ore, trioxide and metal, has been producing far less of its own material in recent years and now relies heavily on imported material, especially from China.

Most Western producing countries have high import tariff barriers, but in times of shortage these countries become net importers from China and other producing countries. Periods of shortage are, however, comparatively short-lived. This situation may change as the major

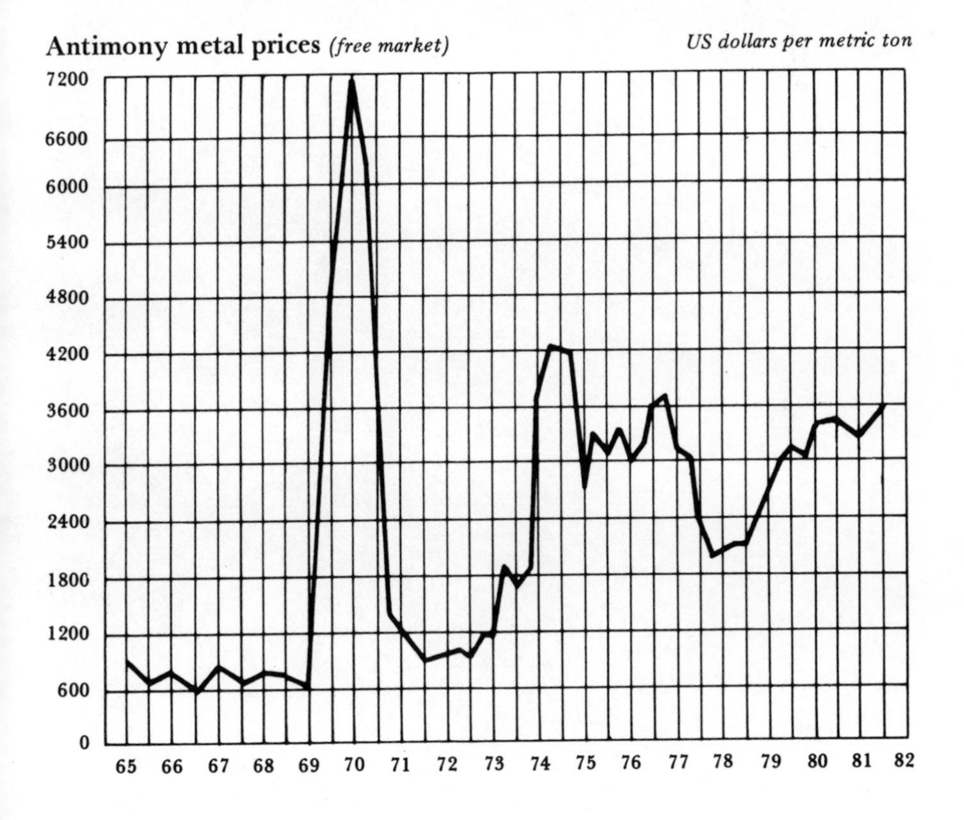

Western producers become less interested in the metal market. Prices are recorded of deals between the merchants operating in the world's major financial areas, particularly in Europe. The Western producers fix a price for domestic sales, but these official prices are greatly influenced by free market trends and may frequently be adjusted.

Scrap recovery

Very little scrap antimony is traded, but there is a comparatively large market in antimonial lead.

Arsenic

Arsenic (trioxide) production, 1979 *thousand metric tons*

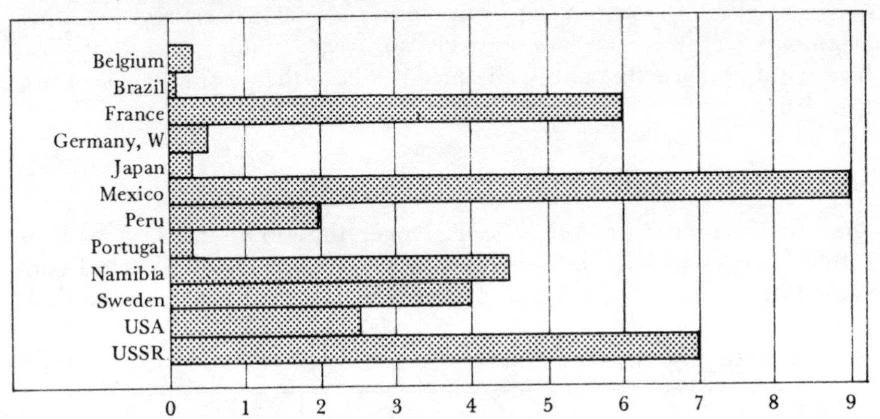

Total world production 36,500 metric tons

Grades available

Elemental arsenic is not strictly a metal in the chemical sense but is referred to as a metal to distinguish it from the oxide in which form it is most commonly distributed. The pure metal is traded in the form of irregularly shaped lumps with a purity of 97 per cent minimum packed in airtight and water-tight drums.

Pure arsenic is rarely traded by merchants but they are a little more active in the arsenic oxide market. The oxide is traded as white powder with a minimum purity of 99 per cent. It is, of course, extremely poisonous and is packed in sealed drums.

Production method

Arsenic is found in many different types of minerals which usually contain other metals particularly lead, copper, zinc, nickel and gold and it is also found together with iron and known as pyrite.

The arsenic is produced from these minerals either as a by-product or co-product of the metal with which it is associated.

In most smelting methods the arsenic is collected as a flue dust in the metal refining process. It is then added to pyrite or galena and heated until the arsenic comes off as a gas which is then collected.

Major uses

About 97 per cent of arsenic production is in the form of oxide which is mainly used to make a large variety of pesticides. Other uses are as a decolourizer in the glass industry and in metal refining.

Pure arsenic is used as an addition to lead in alloys to increase hardness and to copper where it increases corrision resistance. Copper arsenic pipes were used extensively in steam train locomotive fireboxes but this use has almost disappeared with the passing of the steam train era.

Main market features

Arsenic production is widely spread over the globe and the oxide is rather inexpensive. In the last few years, however, environmental considerations have discouraged production as well as consumption.

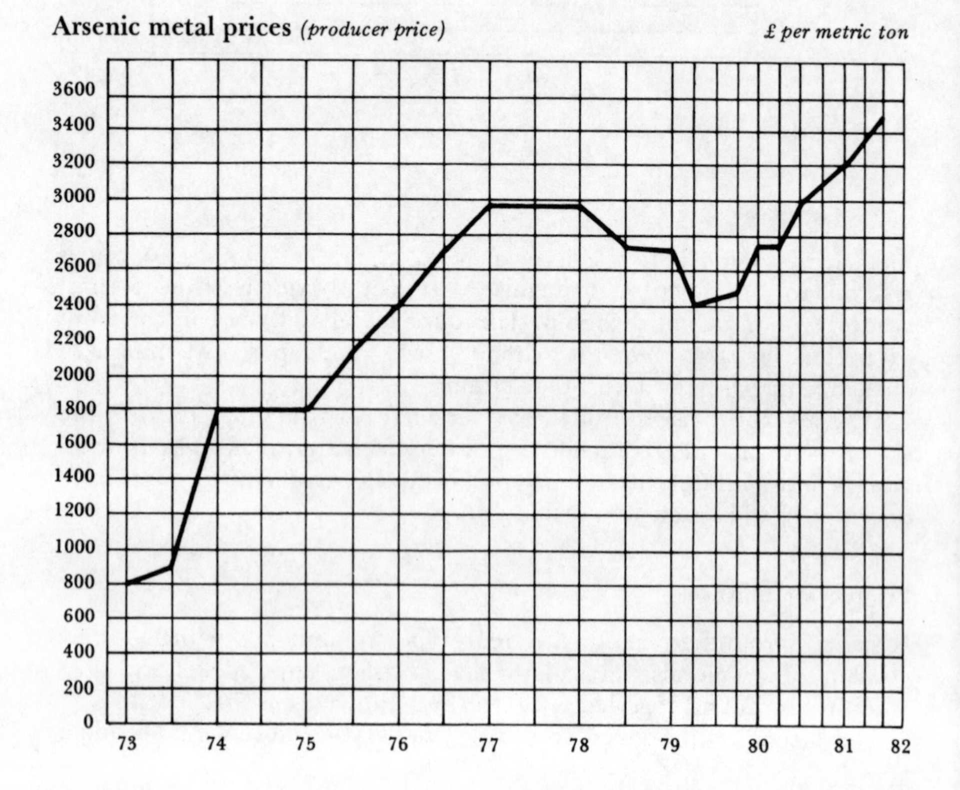

Sweden, traditionally the largest producer, effectively controlled the world's arsenic prices but production in Sweden has now fallen so low that producer prices have become less important over recent years. In this same period production has fallen faster than consumption, which has produced the surprising anomaly of a shortage in this highly toxic material. This situation is not likely to continue, however, as there is no theoretical difficulty in producing arsenic, especially in developing countries where pollution controls are less stringent.

Pure arsenic is much more expensive than the arsenic oxide, but most oxide producers are not interested in producing the elemental form from the oxide. This has led to short periods of active free market trading in times of shortage but demand is somewhat erratic and no sustained period of shortage is possible because so little of the oxide is used to produce the metal.

Method of marketing and pricing

Pure arsenic metal prices are closely controlled by producers but a free market does emerge at premium prices in short periods of high demand. Merchants are only occasionally involved in the oxide market but trade is now highly competitive between producers.

Beryllium

Beryllium ore production (metal content), 1977 *short tons*

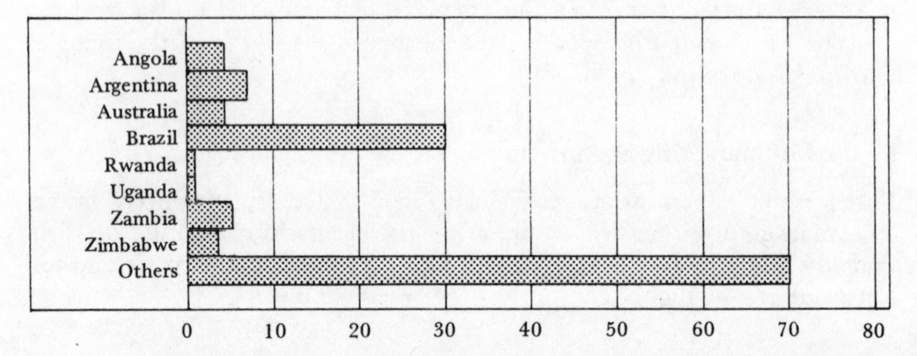

Total world production 125 short tons

Grades available

Pure beryllium metal is rarely traded. Apart from being an extremely toxic metal, which makes handling very dangerous, it is only used commercially in the form of an alloy with other metals. It is therefore safer and easier to market beryllium in the form of a master alloy (an alloy with a high beryllium content compared with its content in the final form) which can then be diluted in the melt. The most common of these master alloys is beryllium-copper with a content of up to 5 per cent beryllium.

Production method

There are several ways of extracting beryllium oxide from its ore, beryl. A common method is to fuse beryl with potassium carbonate. It is then treated with sulphuric acid and the silica is filtered off with water. The liquor containing beryllium is then poured into hot ammonium carbonate and beryllium carbonate is precipitated. When this is redissolved in ammonium carbonate, beryllium oxide is precipitated. The metal is produced from the oxide by reduction in an electric furnace.

Major uses

Most beryllium is used in the form of beryllium-copper, a very important copper alloy used to make marine propellors, springs, aircraft engine components and electrical contacts. It is also used to make non-spark producing tools for use in coal mines and oil refineries, where there is a danger due to inflammable gases. There are minor applications for other beryllium alloys such as beryllium-nickel, and some pure beryllium is used, mainly in nuclear energy plants.

Main market features

Although beryllium ore is produced in many countries, the production of the metal is in very few hands and these are mainly in the US. These producers effectively control the market. This situation is understandable since there are dangers involved in producing the metal due to toxicity and, in addition, the market is small. Producers are rarely asked to supply pure beryllium metal, but conduct most of their business in master alloys.

There is almost no merchant activity in this metal.

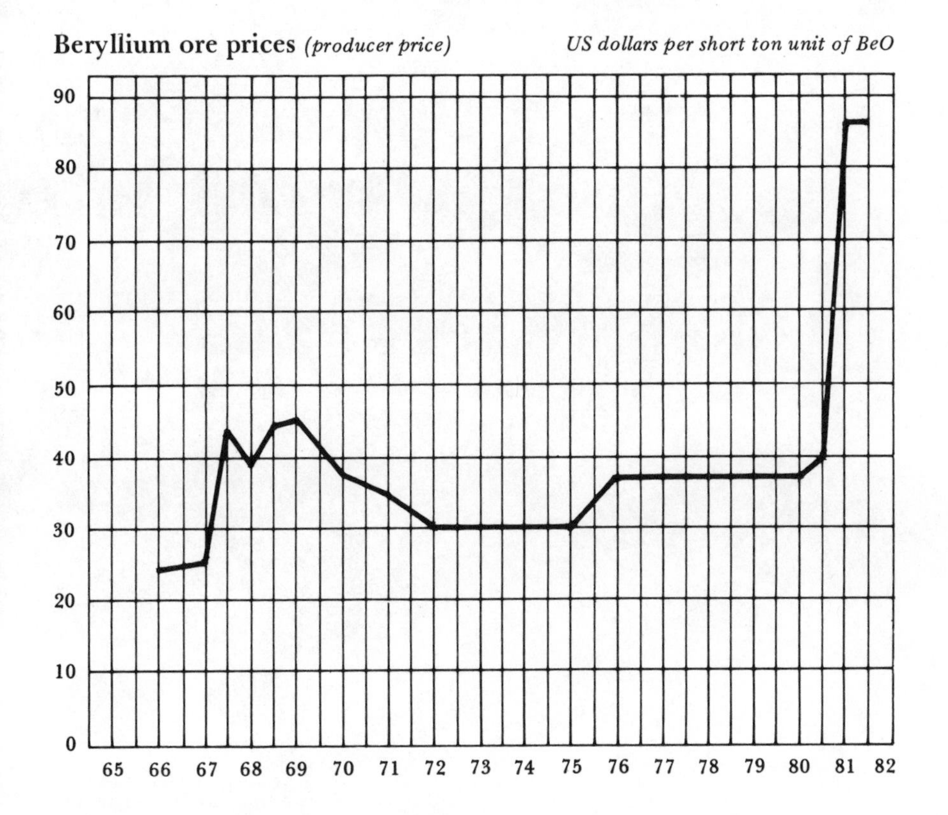

Method of marketing and pricing

The main producers have been successful in controlling prices and there is practically no free market outside this system.

Bismuth

Bismuth metal production, 1979 *hundred metric tons*

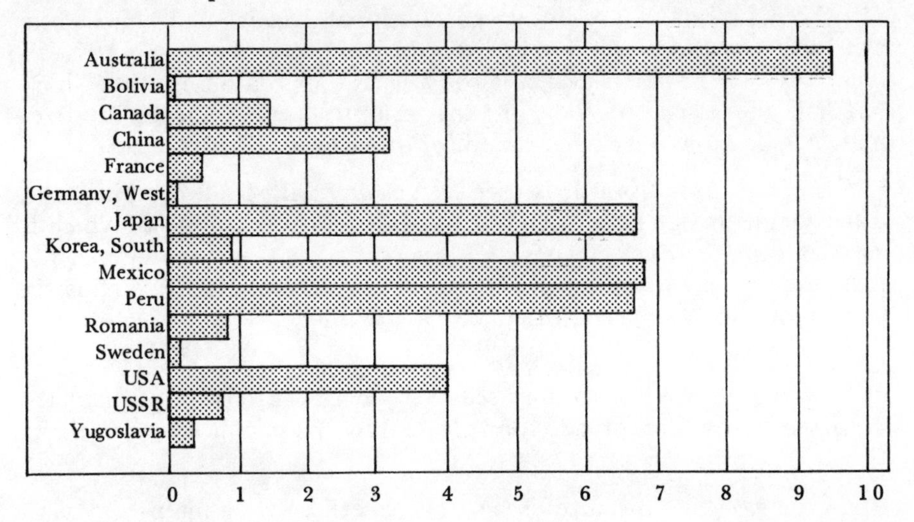

Total world production 4,100 metric tons

Grades available

Bismuth is marketed in the form of ingots usually weighing 10 kilos each and packed in wooden boxes each weighing 100 kilos, or in the form of needles or granules weighing between 3 and 5 grams each and packed in 100 kilo steel drums.

Purity is 99.95 per cent minimum up to 99.99 per cent minimum. Some small producers market bismuth in a low refined state (around 96 per cent purity) known as bismuth bullion, usually contaminated with lead. Consumers of the metal who produce pharmaceutical products may demand a low tolerance of trace elements.

Production method

Usually as a by-product of lead or zinc manufacture but also as a by-product of copper, silver, gold and tin refining. The bismuth rich concentrates, usually sulphides, are oxidized in a simple and inexpensive process to the metal. It is then upgraded by skimming and recrystallization and treatment with flucculents to produce metal with a purity of 99.99 per cent suitable for most industrial and pharmaceutical uses.

Major uses

a) *Pharmaceutical* Some salts of bismuth have been found beneficial in the treatment of indigestion and other minor ailments of the alimentary canal and these account for over half the world's consumption. The French, however, who have traditionally consumed almost one third of the world's production for this purpose, decided in 1974 to make bismuth products available to patients only on prescription which has had the effect of reducing that country's consumption by over half. Applications have been found for other bismuth salts in cosmetics.

b) *Fusible Alloys* Bismuth is used as an alloy with lead, cadmium and other metals in the manufacture of fusible alloys, the bulk of which is used as jigs for making parts in the aircraft industry, and also in the manufacture of optical lenses. Other uses are found for these alloys in press moulding machines, the dyeing of textiles, fire sprinkler systems and solders.

c) *Catalysts* Bismuth salts are used in a range of catalysts, particularly in the manufacture of acrylonitrile, a raw material for a man-made fibre.

d) *Alloying agent* Bismuth is added to certain aluminium alloys to improve machinability and also to other metals for specialized uses.

Main market features

Consumption of bismuth has been subject to large fluctuations. The popularity of bismuth as a catalyst in the 1960s and the legislation concerning medical products in France, together with the fortunes and misfortunes of the world's aircraft industry, have all contributed to make the bismuth market one of the world's most volatile. The price volatility itself has contributed to nervousness amongst potential consumers, but bismuth has some unique physical characteristics which must find a continuing interest in industry, especially at its current very low price.

In 1974 France consumed 1,259 metric tons, almost one third of the world's bismuth production. By 1977 consumption in France had fallen to 644 tons and by 1978 to only 346 tons. This fall was a direct consequence of action taken by the French medical authorities to curb the use of bismuth in pharmaceutical products after doubts had been raised concerning side effects suffered by some patients who took bismuth preparations in large doses.

One of the largest bismuth producers in recent years is Australia, where it is produced as a co-product with other metals. It is unlikely that any shortage of bismuth could occur for several years.

There is no official market for bismuth. The major producers fix an official price, but in times of shortage the material is rationed to consumers and the official price is often raised. In times of glut the producers offer discounts on the official price.

It is true to say that since the late 1960s the major producers have lost control of the pricing of this metal and free market forces have taken over. Although perhaps only 25 per cent of the world's production is distributed by merchants, it is their transactions that give the best guide to the price movements.

The merchants who specialize in bismuth, trading mainly in New York, London, Brussels and Dusseldorf, trade the metal throughout the working day, buying surplus stocks from producers and distributing to consumers. These merchants often speculate by holding stocks or 'going short' of the metal and therefore often trade between themselves. They trade in minimum one metric ton lots, but average transactions would be five metric tons.

Bismuth metal prices *(free market)* *US dollars per lb*

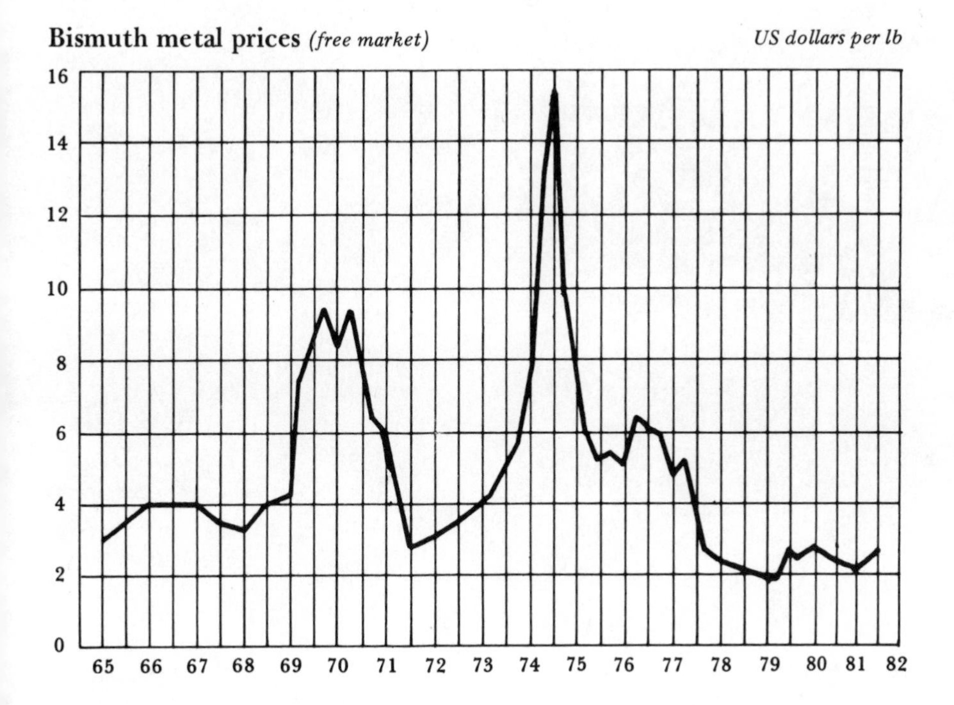

There is very little scrap trade in bismuth, except for some adulterated alloys that may need extensive refining.

Estimated world reserves

93,000 metric tons.

Cadmium

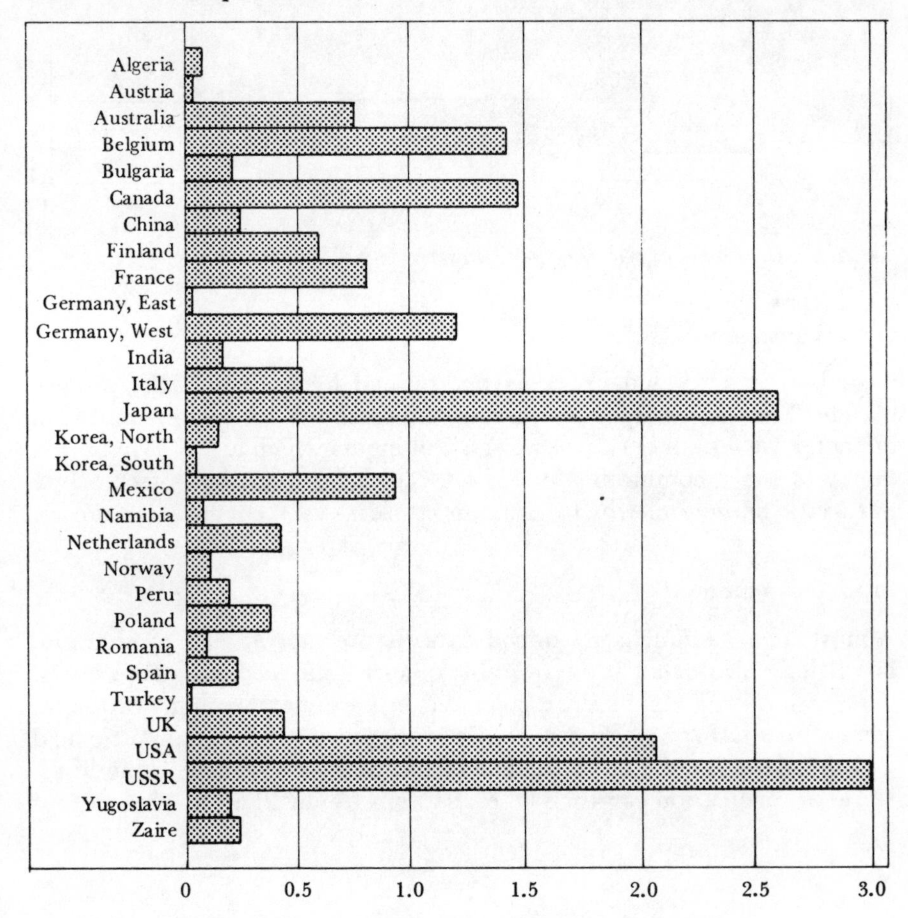

Total world production 19,044 metric tons

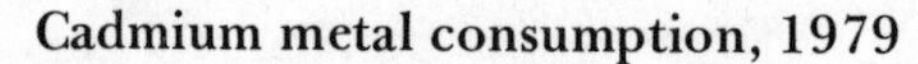

Cadmium metal consumption, 1979

thousand metric tons

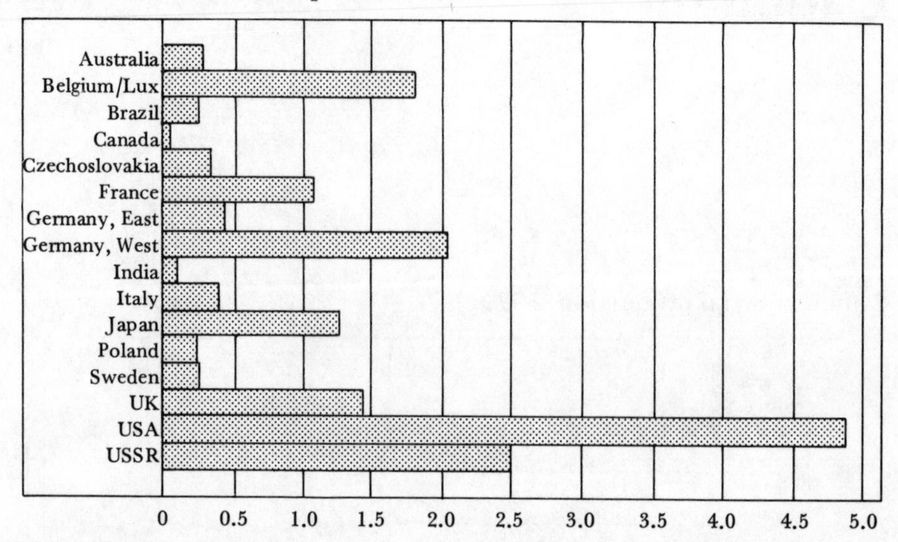

Total world consumption 18,493 metric tons

Grades available

Cadmium is marketed in forms determined by its use: balls approximately 2 ins in diameter for plating, sticks about 10 ins long and ½ in diameter for use in chemical works, and ingots which are used in alloys. Purity is most commonly 99.95 per cent minimum, but often 99.99 per cent minimum which is required for some chemical purposes.

Production method

Almost all cadmium is produced as a by-product of zinc production, but it may also occur in recoverable quantities in lead and copper ores.

Cadmium rich residues are collected in base metal refining processes usually in the form of flue dusts. It is then dissolved in sulphuric acid and the metal is either separated electrolitically or by chemical methods. Further purification can then be carried out by distillation.

Major uses

Cadmium has four major uses – in plating, pigments, polyvinyl chloride stabilizers and in alloys. Plating is the most established use. Cadmium plating of steel products is carried out to ensure a firmer base for another plated metal, like chrome, which would be electroplated over the cadmium to counteract certain corrosive elements, such as sea water. Mechanical and electronic equipment used in coastal or marine work often requires cadmium plating.

Cadmium salts which are made directly from the metal are used to make pigments for ceramics and plastics and are used as stabilizers in the manufacture of PVC. Cadmium is highly toxic and this prevents it being used in any material that comes into contact with foodstuffs.

The metal is also used in combination with tin, lead and other metals to make a variety of solders, print metal and fusible alloys used in sprinkler systems and in fusible jigs.

Main market features

Cadmium production is as widely spread as zinc production and no single producer is in a position to control the market. Like many similar metals the North American producers publish a producers' price for home buyers and for exports. Most other producers, however, don't seem to set a high priority on controlling cadmium prices, and sell at the best price they can to international merchants who specialize in the trading of minor metals.

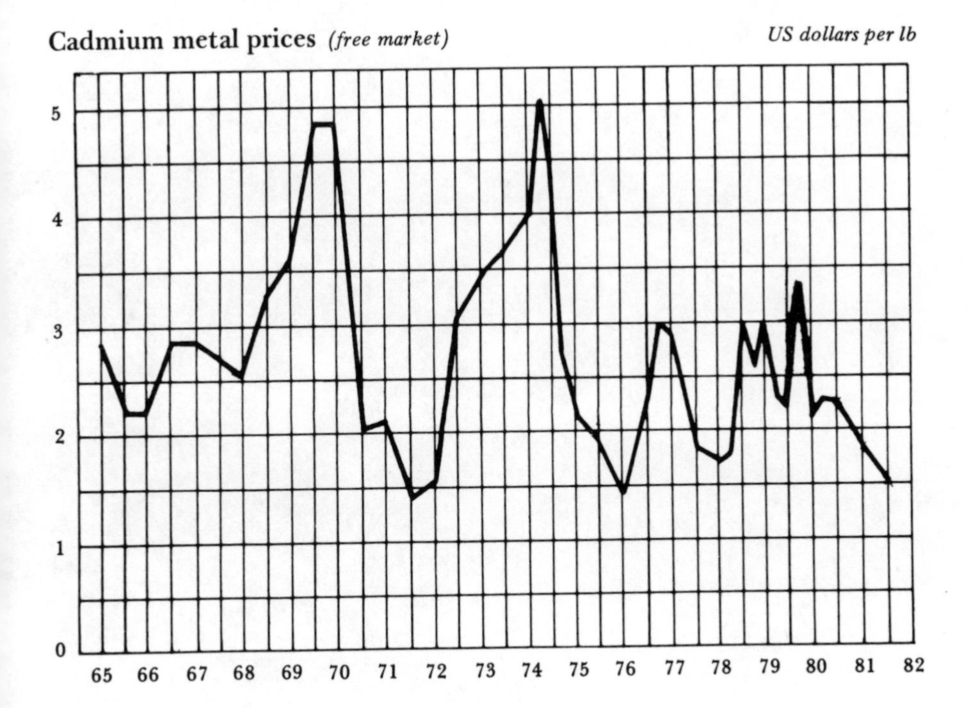

Cadmium metal prices *(free market)*

Known world reserves

Approximately 550,000 metric tons.

Method of marketing and pricing

Cadmium is a favoured trading item for metal merchants who trade on a free market price. The cadmium market is fairly small and very volatile, due to merchant manipulation. The fact that cadmium is a by-product of zinc production means that more metal cannot be produced simply because demand rises.

The market has also attracted the attention of non-trade speculators, which has further distorted the normal rules of supply and demand.

Producers and consumers alike are increasingly concerned with pollution problems associated with cadmium as it is a highly toxic metal. This may tend to reduce activity in this market.

Cerium

Cerium (monazite ore) production, 1975

thousand short tons

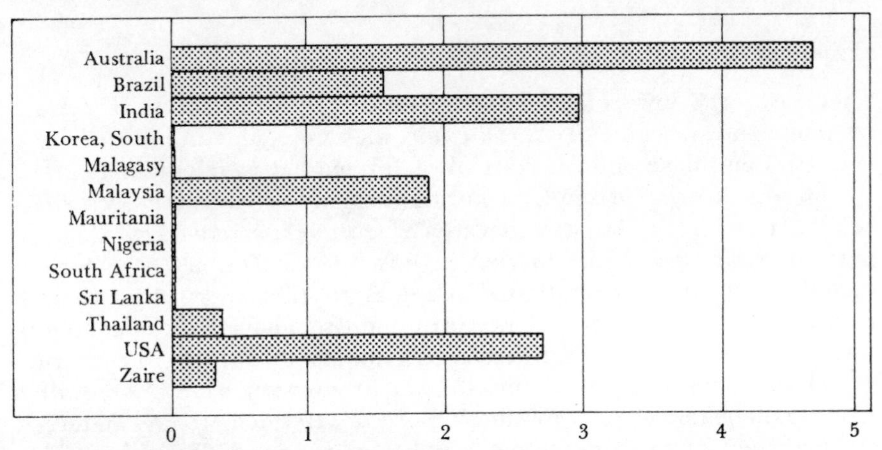

Total world production 11,895 short tons

Grades available

Pure cerium metal is of no interest to the metal dealer and is only produced for laboratory uses and research purposes.

Mischmetal, however, a mixture of rare earth metals containing about 50 per cent cerium, is regularly traded in the form of small ingots, pellets and notched bars.

Production methods

There are several ores containing cerium and other rare earth elements, the most common of which is monazite. It is also found in commercially recoverable quantities in xenatime, bastnaesite and loparite. Monazite is found in a beach sand ore and the rare earths are recovered here as a by-product of titanium production. Rare earth ores are separated from the beach sand by magnetic and electrostatic means. They are concentrated by flotation and the metal (mischmetal) is produced by electrolysis of the rare earth chlorides.

Major uses

Cerium compounds, particularly the oxide, have several important uses as filters and de-colourizers in the glass industry and as catalysts in petrochemical production.

Mischmetal alloyed with iron is used as lighter flints, but its main metallurgical use is as an addition together with magnesium to certain cast irons known as nodularized irons. It is also added to certain types of steels to increase strength and ductility.

Main market features

There are very few companies engaged in the manufacture of cerium products and mischmetal. These companies are mainly in industrialized countries and have enough control of the market to maintain an effective producer price in times of normal demand. There have been a few brief periods in the last two decades when unexpectedly high demand has led to an active free market in mischmetal. During these periods major producers have rationed material to their regular customers. Consumers, especially in the cast iron industry, have then appeared as buyers often at quite substantial premiums to the producer price.

Temporary shortages of mischmetal production have also resulted from inadequate supplies of suitable ore or intermediate raw materials.

The effect of small mischmetal additions to iron and steel is a rather complicated and interesting branch of metallurgy and new uses for this compound are expected to increase demand in this field.

Known world reserves

Cerium containing ores are 10 million tons of rare earth oxide contained, about half of which is estimated to be in the US.

Method of marketing and pricing

As has been mentioned, a few producers of cerium products operate in industrialized countries where they sell direct to consumers, especially on their domestic market, at fixed producer prices. During the comparatively short-lived periods of shortage, however, mischmetal has attracted attention from international merchants operating at premium levels.

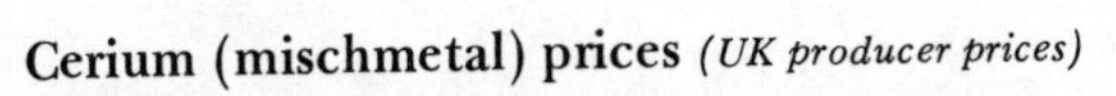
Cerium (mischmetal) prices (UK producer prices)

£ per kilo

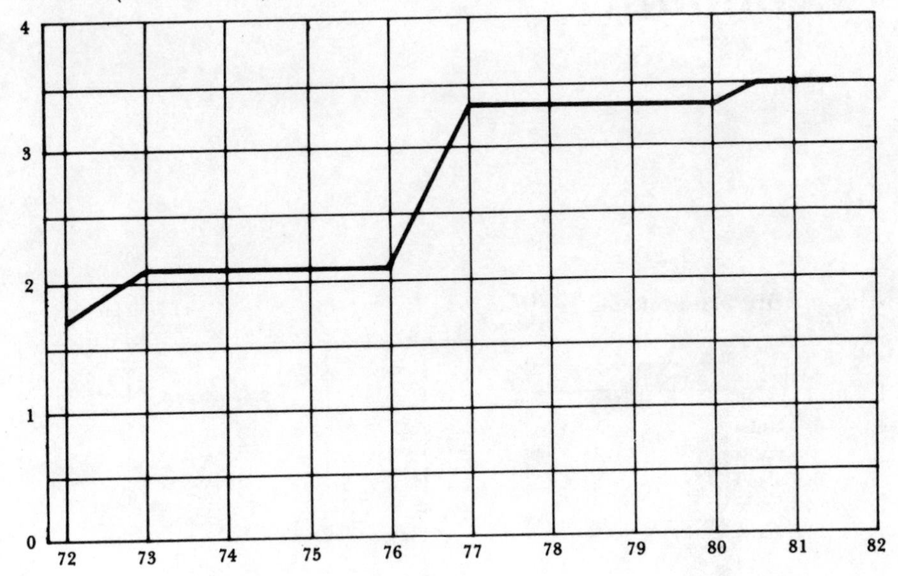
4
3
2
1
0
72
73
74
75
76
77
78
79
80
81
82

Chromium

Chrome ore production, 1979

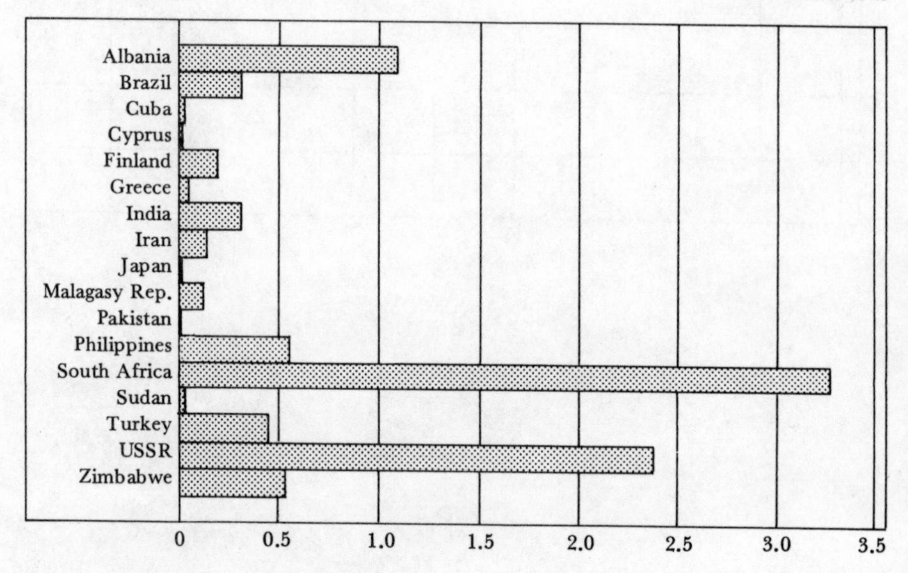

Total world production 9,600,000 metric tons

Grades available

Only a small part of chrome produced is converted into chromium metal. The bulk of chromium is used in the form of its ore (chromite) or ferro-chrome. Chromite ore is simply mined, cleaned, crushed and graded. Ferro-chrome, the alloy of chrome and iron, is produced in several grades. Charge-chrome (containing about 55 per cent chrome), high carbon ferro-chrome (5-8 per cent carbon); and low carbon (0.1 per cent carbon maximum). Of these three, the cheapest is charge-chrome. Another inexpensive chrome containing alloy is ferro-silicon chrome.

Chromite is invariably sold in bulk but ferro-chrome is sold in a lumpy form and may be transported in bulk or packed in steel drums.

Production methods

Chromite is melted in an arc-furnace together with iron to produce ferro-chrome. Chromium metal is produced either by electrolysis of a chromium solution or by alumino thermic reduction, whereby chromium oxide is mixed with aluminium powder and ignited, thus reducing the oxide to chromium metal.

Major uses

About 20 per cent of the chromite mined is used directly in refractories, foundry mould sand and other non-metallurgical uses. Its other two major uses are in alloy form in steels and for plating. Chrome is added, in the form of ferro-chrome, to steel to produce alloys including stainless steel. There is a large range of very useful chrome steels but by far the most common of these is normal stainless steel containing both chrome and nickel.

Pure chromium metal is used to manufacture many types of non-ferrous alloys and super alloys, used in magnets amongst other things. Chromium is plated on to such items as automobile trim and cutlery from a bath of chromic acid which is made directly from chromite.

Method of marketing and pricing

World trade in chromium can be confused by the complications of tariff barriers and quota systems in individual markets. However, the price trend is set by South African producers and followed competitively by Russian exports. Very recently the USSR has been a little more erratic in its sales policy and has occasionally withdrawn as sellers to the market. This reflects the general trend of the USSR consuming more of its own metal production. Most business is conducted by producers and their agents but there is an extremely active free market conducted by large merchants utilizing particularly Russian supplies and material from the smaller producing countries. Business is highly competitive mainly because the important trade is conducted with very large steel mills which usually place business on a long-term basis. There are many nationally based merchants who trade on a smaller scale with small specialist steel works and foundries.

Main market features

Perhaps the most outstanding feature of the chromium market is the fact that the US, the world's biggest end-user of chrome, has no indigenous production.

Chromium is one of the world's most important industrial metals and recent events have once more underlined the potential vulnerability of the US chromium consumer to foreign political events.

South Africa and the USSR produce between them about half the world's chromium in the form of chromite ore and ferro-chrome. Together they represent the bulk of the world's exports.

It can be understood that the US is not too happy about being dependent on either the USSR or South Africa for the supply of such an essential raw material.

Zimbabwe is a major producer of chromium, accounting for some 8 per cent of world production. Much of this material has been exported by various devious means during the period of UN sanctions. The future of that country appears healthy but should a major upheaval occur chrome supply could again be disrupted.

Another important feature of the chrome market is the technical innovation in the production of stainless steel. The comparatively new AOD (Argon-Oxygen Decarburising Vessel) has been found to be the most efficient method of producing stainless steel and among its technical advantages is its ability to utilize lower quality raw materials including chrome.

In the past, ferro-chrome producers were encouraged to reduce the carbon content in their product, an extremely expensive operation. With the advent of the AOD, carbon content can be substantially reduced in the steel making process. Ferro-chrome producers, aware of this change, have swung over on a large scale to the production of charge chrome with a chrome content of about 55 per cent as opposed to the 70 per cent chrome content in standard, medium, and low carbon ferro-chrome. The carbon content of charge chrome

Low carbon ferro-chromium prices *(free market)*

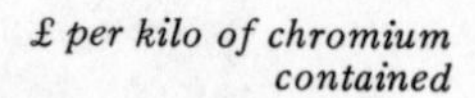

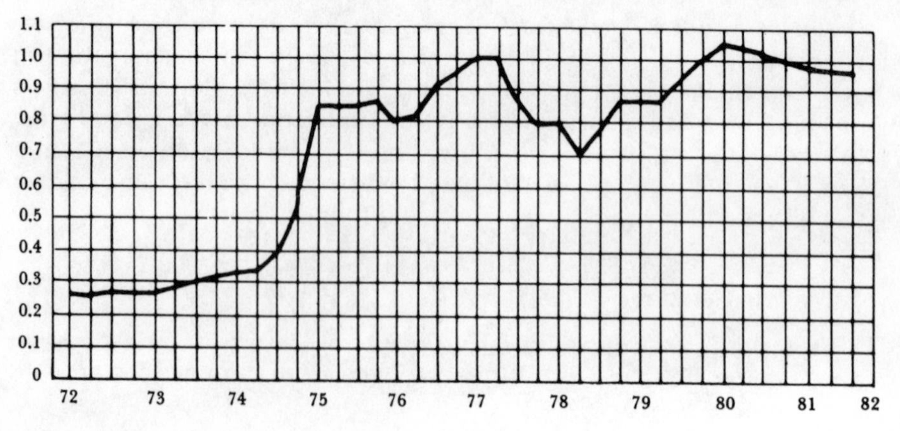

can be as high as 6-8 per cent without worrying stainless steel producers using AODs.

Due to the strategic importance of chromium supplies, several Western governments are considering the stockpiling of this metal.

Known world reserves

Approximately 775 million metric tons of chromite ore. 575 million metric tons are estimated to be in South Africa.

Cobalt

Cobalt metal production, 1978

thousand metric tons

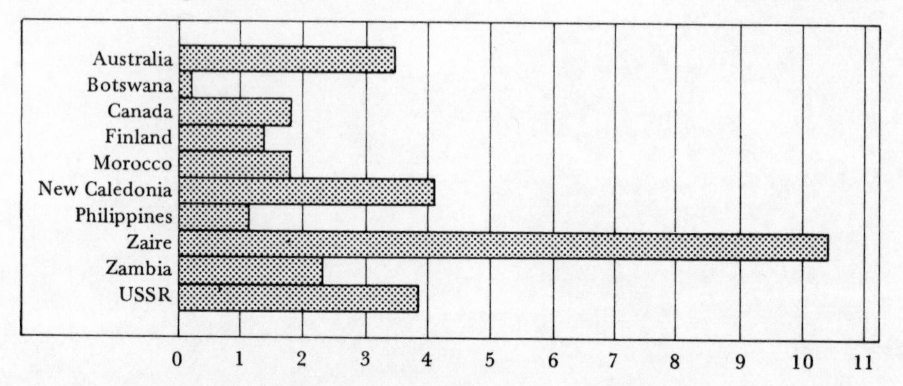

Total world production 30,300 metric tons

Grades available

The most commonly traded form of cobalt is broken cathodes, small flat pieces of the metal produced by crushing a cathode of the brittle pure metal. Granules are a less common form of the pure metal but both are normally distributed with a purity of 99.6 per cent minimum.

Production method

Cobalt is almost universally produced as a by-product, or co-product, of either nickel or copper, except in Morocco where the metal is mined on its own with copper as the by-product. The cobalt rich concentrates are normally collected during the electrolysis of the base metal concerned and is itself produced by electrolysis.

Major uses

The two major uses for this metal are in what is termed super alloys used extensively in the aircraft industry (it is impossible at present to make a modern jet engine without the use of cobalt alloys) and in turbines. Super alloys are usually high alloy steels but may contain only other non-ferrous metals. The second major use is again in alloys, consisting of cobalt and other non-ferrous metals used to make magnets and other electromagnetic components in the electronics industry.

Cobalt alloys are extremely useful in applications where strength at high temperatures is required. There are certain important applications for cobalt salts, particularly the oxide, which is used as a catalyst in the petrochemical industry and as pigment in ceramics and enamels.

There is a very minor use for cobalt isotopes in the treatment of cancer. Pure cobalt powder has an application in sintered products such as machine tools and ammunition when combined with other refractory metals.

Main market features

Many of cobalt's applications are in highly strategic areas, such as its use in aircraft engines and fuselages. It is interesting to note that although the US consumes over one third of the world's production, directly or indirectly, it has almost no production of its own.

Before 1978 Zaire accounted for about 65 per cent of the world's cobalt production. However, in that year Shaba, the important mining province of Zaire, was raided by a rebel army which had the effect of disrupting production for some months and driving away most of the skilled white engineers who worked on cobalt production. This caused cobalt prices to increase by some 500 per cent in a few months and, although prices fell again when production resumed once more, at a reduced output, there is still a shortage of supply.

High prices have stimulated increased production from all the other major producers and stimulated some new mining ventures which eventually alleviated the shortage.

Zaire and Zambia (the second largest African producer) have considerable political and economic difficulties, and the possibility of further supply disruption cannot be ruled out. On the other hand, both countries have a great incentive to produce as much cobalt as possible and have announced plans to increase production considerably.

High prices have encouraged consumers to recycle as much scrap material as possible and to find suitable substitutes.

Known world reserves

Approximately 1,100,000 tons of which 700,000 tons are estimated to be in Zaire.

Method of marketing and pricing

Until the disruption of supplies from Zaire, 90 per cent of all cobalt produced was marketed on a producer price basis. The sudden price rise in 1978 and the subsequent rapid fall, however, encouraged a very active free market in the metal. Producers have been slow to reduce their prices and now consumers buy up to 30 per cent of their requirements from the 'free market' or at 'free market' prices from smaller producers. The United States authorities have recently bought several hundred tons of the metal (directly from major producers). These quantities are unlikely to affect the market unduly, however, as there are now stocks amounting to over 7,000 tons in producer hands.

Cobalt metal prices *(free market)* *US dollars per lb*

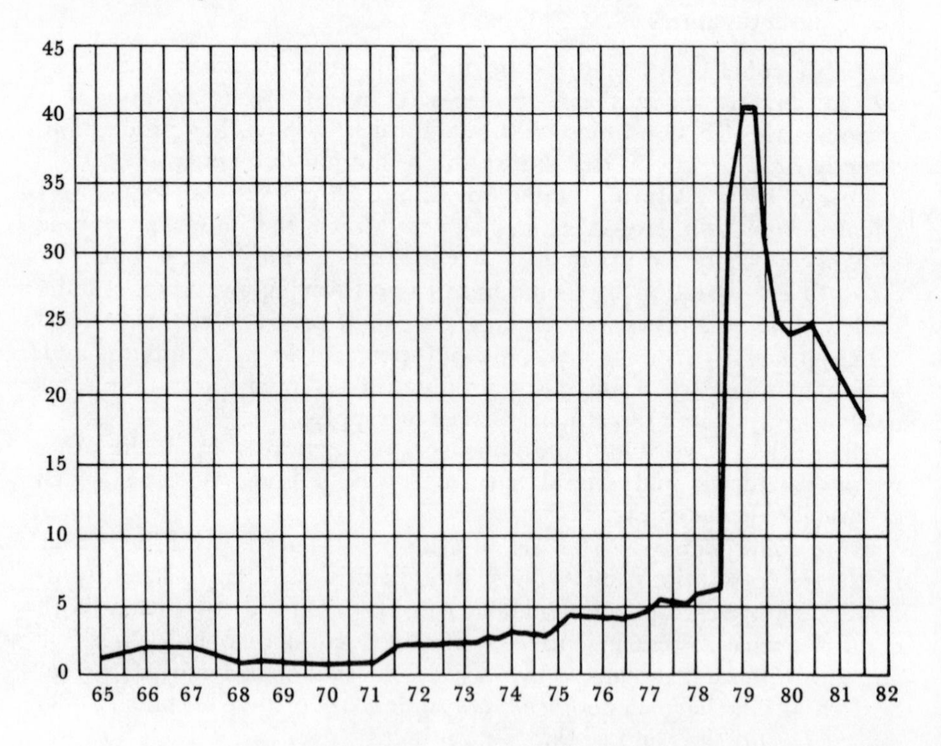

Columbium or Niobium

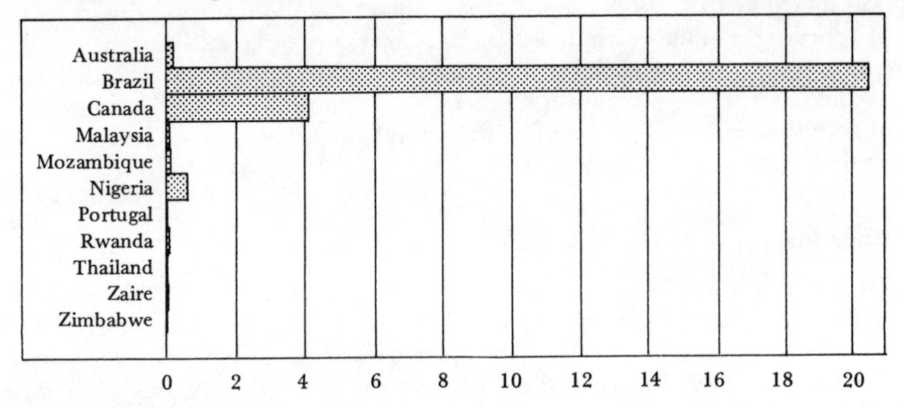

Total world production 25,000 metric tons

Grades available

Pure columbium metal is a rather rare item accounting only for a tiny fraction of total columbium consumption. Its most commonly traded form is ferro-columbium (its alloy with iron) containing from 60-70 per cent columbium. In this form it is sold as irregular lumps packed in drums.

Production method

Columbium is extracted from its naturally occurring ore pyrochlore and from the ore columbite which strangely contains both columbium pentoxide and tantalum pentoxide. If the tantalum content exceeds the columbium content the same basic mineral is known as tantalite. The ore is concentrated by a combination of wet gravity, magnetic and electrostatic processes.

In columbite and tantalite, columbium oxide is separated from the tantalum oxide first by conversion to the fluorides then by a solvent extraction method. The columbium is then reconverted to the oxide by calcining.

Ferro-columbium is produced by the thermic method, that is burning the pentoxide with aluminium powder in the presence of iron powder. Columbium metal can be produced by reducing columbium chloride with magnesium metal.

Major uses

Almost all columbium is used in the form of ferro-columbium and more rarely in the form of the pentoxide in the manufacture of alloy steels. These are in turn used for structural purposes in buildings, bridges, etc, for heavy mining equipment such as rock cutters and for machine components where shock resistance is required. A recent growth area for columbium consumption is in the manufacture of gas and oil pipeline steels. Minor uses for the metal occur in the nuclear energy and electronics industries.

Main market features

Brazil is responsible for producing about 75 per cent of the world's supply of this metal and holds a similar proportion of the world's known reserves. Columbium producers have so far had a very responsible attitude to pricing, with prices rising steadily in line with the growth of consumption and inflation. The effects of the 1974 oil price rise has, however, greatly encouraged the construction of oil pipelines producing an unforeseen demand for columbium. Other metals used for similar purposes including pipeline steels, vanadium and particularly molybdenum, have shown marked price rises which may justify a price rise in columbium. Brazil is in a uniquely powerful position to raise prices if demand justifies such a move. But the competition from vanadium, which can be substituted for columbium in many alloys, needs to be remembered.

Known world reserves

5,100,000 metric tons of columbium (metal content) of which approximately 4 million metric tons are estimated to be in Brazil.

Method of marketing and pricing

Columbium has not been a particularly popular item with metal merchants in the past due mainly to the efficient and flexible pricing policy of main producers who are able to adjust prices and stocks according to demand. They have none of the problems of molybdenum producers, for instance, whose metal is largely a by-product material.

Most producers either sell directly to consumers or have local agents to market their product. International merchant activity is only apparent during temporary shortages of material.

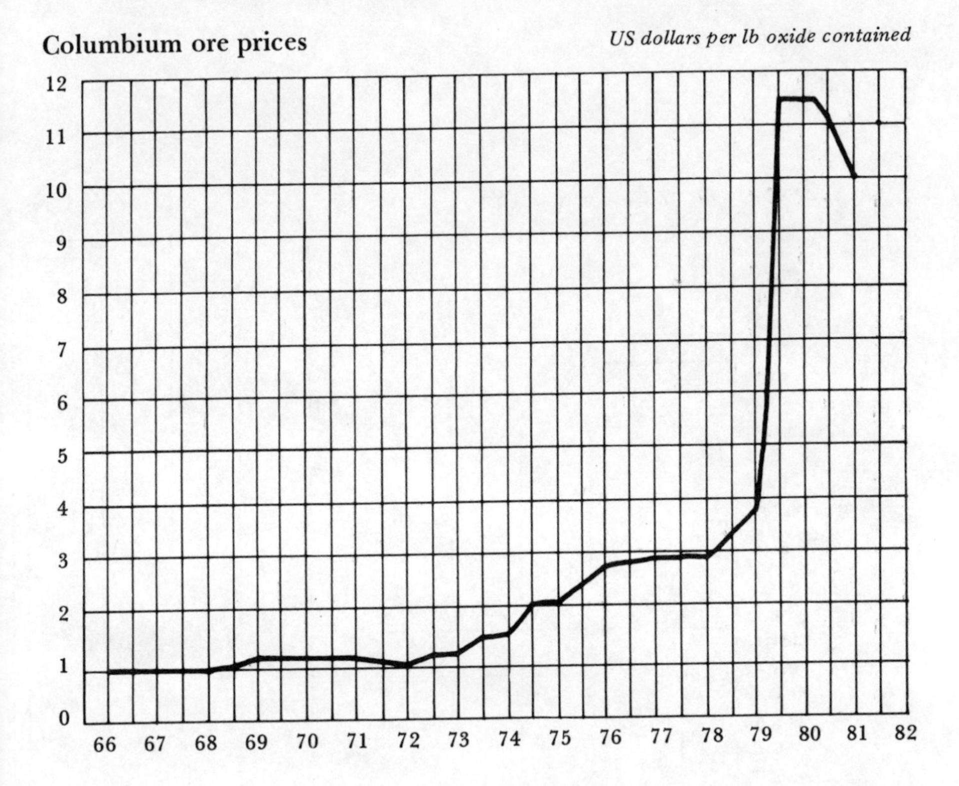

Copper

Copper refined production, 1980

million metric tons

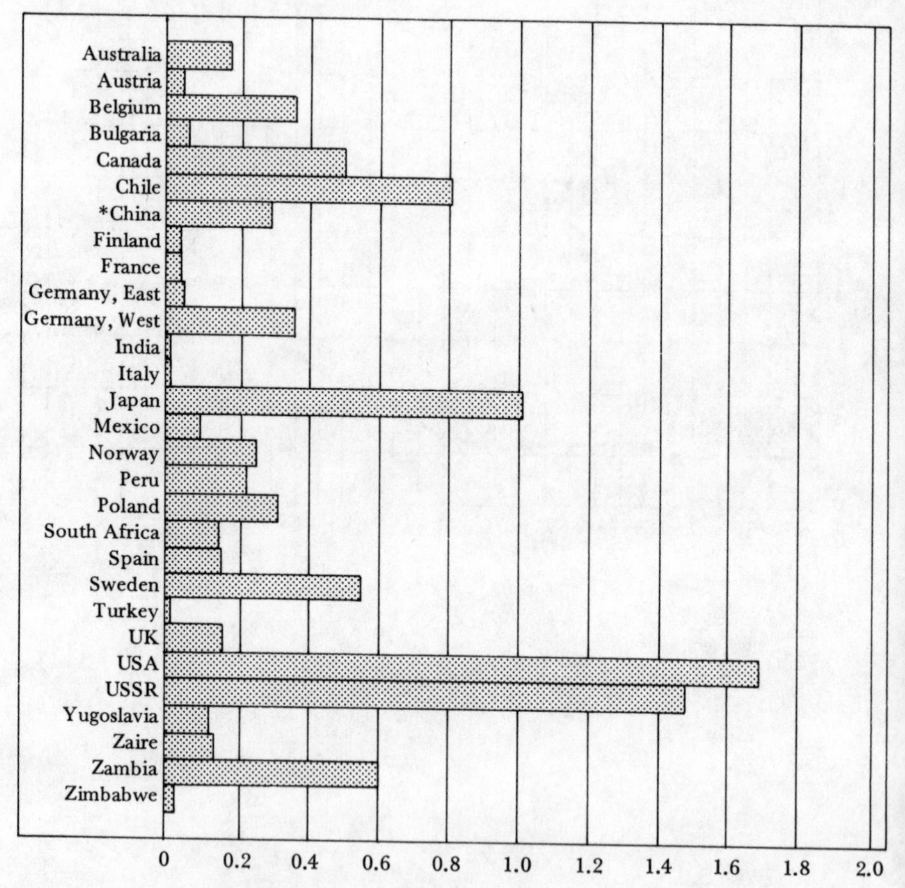

Total world production 9,373,400 metric tons

** & other Asia*

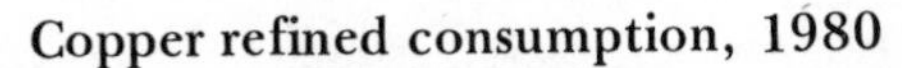

million metric tons

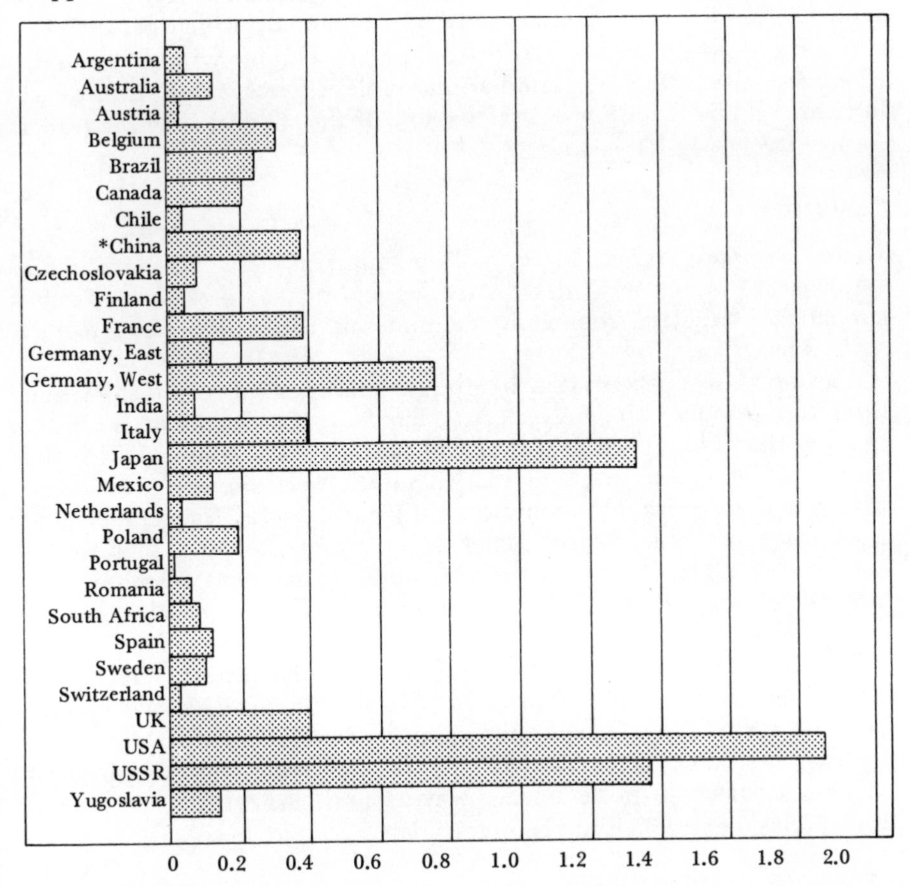

Total world consumption 9,528,200 metric tons **& other Asia*

Grades available

Pure copper is most commonly traded in two forms, cathodes and wirebars.

A cathode is the pure copper slab which results from the electrolytic refining process. Its usual dimension is about 36 ins x 36 ins x ½ in. It has a bubbly surface and a minimum purity of 99.9 per cent copper. This purity includes a small silver content.

Wirebars which are generally cast from cathodes are the elongated ingots used as a starting rod for drawing copper wire. Its standard weight is between 90-125 kg, again with a purity of 99.9 per cent copper plus silver.

Many producers offer grades with certain chemical characteristics such as HCFR (High Conductivity Fire Refined) which is used for applications where high electrical or thermal conductivity is required. Producers also offer fabricated forms such as wires, tubes, sheets etc, but these markets are more the concern of the stockist than the trader or broker.

Production method

There are many types of copper ore but commercially recoverable deposits are either sulphides or, less commonly, oxides. Occasionally, copper is extracted from complex minerals containing other metals such as lead or zinc.

The ores are concentrated by the normal methods of crushing and flotation. Copper salts may be extracted by leaching, that is to say treating the ore with an acid that will preferentially combine with the copper and the resulting copper rich solution can, in turn, be treated to extract the metal. Leaching is particularly useful for refining low grade ores or mine waste. Many copper ores contain other useful non-ferrous metals such as molybdenum, cobalt and selenium and methods to extract these metals in refinable form are incorporated in the copper refining process.

The ores may first be roasted, if the required desulphurization is impossible in the smelting process. The smelter produces an impure form of metal known as blister copper which is cast into large flat ingots. These are used as anodes for the electrolytic refining process which is carried out in the normal way using thin sheets of pure copper as cathodes, onto which the copper is plated.

Major uses

Copper's most familiar uses are as wire for conducting electricity and as its alloy brass for screws and fittings. It is in the electrical industry where most copper is consumed.

Other important uses are as pipes for domestic water supply, in the chemical industry as heat exchangers which include car radiators, refrigerators and industrial cooling systems. The above uses exploit the excellent electrical and thermal conductivity of copper, which are important elements in almost all industries although automobile and electrical cable manufacturing are particularly important consumers.

Another major use is in the manufacture of coinage.

Main market features

The two countries where most copper is consumed, the USA and the

USSR, are also the world's two largest copper producers but whereas the USSR is more or less self-sufficient in its supplies of the metal the USA still remains a substantive net importer. The bulk of imports made by the USA and the rest of the developed world are supplied by Third World countries, many of which belong to CIPEC (the Intergovernmental Council of Copper Exporting Countries). CIPEC has attempted to regulate the supply of copper in such a way as to ensure a sustained high price for the metal produced by its members. Its efforts have been somewhat less than successful.

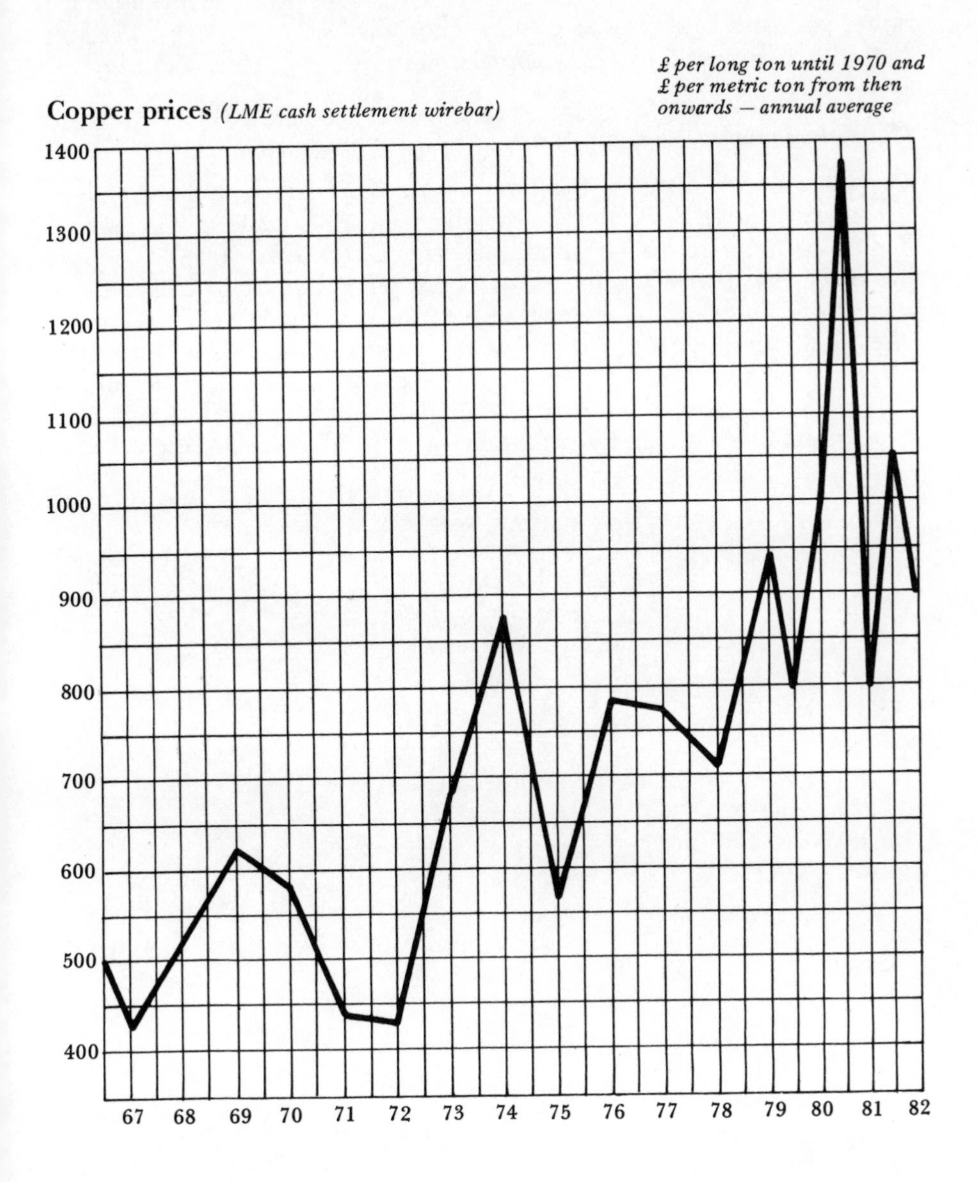

After iron and aluminium, copper is the world's most important industrial metal, a factor which makes control of its supply singularly difficult. Copper exporters, for example, have no control over the circulation of copper scrap, an extremely important element of total copper supplies. Nor do they produce a large enough proportion of copper to exercise complete control of the market. The four most important members of CIPEC, Peru, Chile, Zambia and Zaire, with their vastly differing political orientations, have found it difficult to come to a workable agreement on a tactic for limiting supplies when prices are low, but the present recession may at least give them the incentive for doing so, especially as one effect of the recession has been to reduce output at a greater rate in developed countries.

Method of marketing and pricing

Copper is traded both on the London Metal Exchange and the Comex exchange in New York and almost all the world's trade in the metal is based on the price traded on one or other of these markets. US copper producers have now all but abandoned their producer price system in favour of Comex pricing even though most of them are highly vertically integrated, that is to say they are the largest consumers of their own product. Comex and LME prices are used as a basis for the sale of copper in all stages of its treatment including ores, concentrates, blister copper, cathodes, wire bars, semi-fabricated products and scrap.

Gallium

Gallium production, 1975 *thousand kilos*

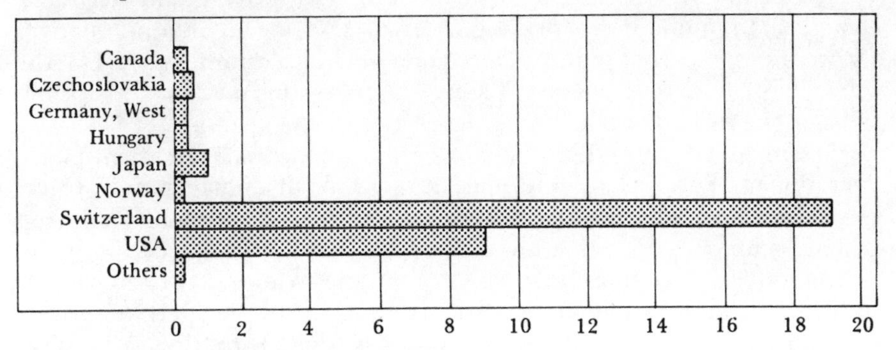

Total world production 30,700 kilos

Grades available

Gallium of a very high purity between 99.9 per cent and 99.9999 per cent is required for most uses. It is usually sold in the form of small ingots.

Production method

All gallium is produced as a by-product of aluminium and zinc production.

It is only extracted from a small proportion of aluminium or zinc ore due to the expense of the process and the low demand in terms of tonnage for gallium. It is extracted from the alumina by precipitation from a caustic solution, then by electrolysis with gallium deposited at the cathode.

Gallium is extracted from zinc concentrates by a simple chemical process and electrolysis, using platinum electrodes.

Major uses

Gallium has two main uses: in electrical semi-conductors and in measuring devices. Its use in semi-conductors represents over 90 per cent of consumption, mainly for diode manufacture.

Main market features

The fact that gallium is a by-product does not mean that the potential for production is limited by the production of the base metal ore from which it is produced, since so little of the main ore is made available for gallium extraction.

Any increase in demand could be met, but time might elapse before plant could be prepared to do this work.

The semi-conductor industry is extremely fickle and regularly changes its requirements for raw material. This means that alternatives for gallium may be found after some technical innovation. On the other hand, it is quite likely that new uses for this metal could arise in some sophisticated application.

No single producer has the power to exercise any great influence over the market, but, as the market is so small and technical, prices remain high to compensate for the stringent level of quality control, and of technical advice that must be offered by any producer.

The price of gallium rose sharply in the winter of 1980-81 due mainly to a somewhat erroneous assumption on the part of private speculators that the gallium market was similar to the germanium market. Prices have since fallen and may continue to do so as production is increased.

Gallium metal prices *(free market)* *US dollars per kilo*

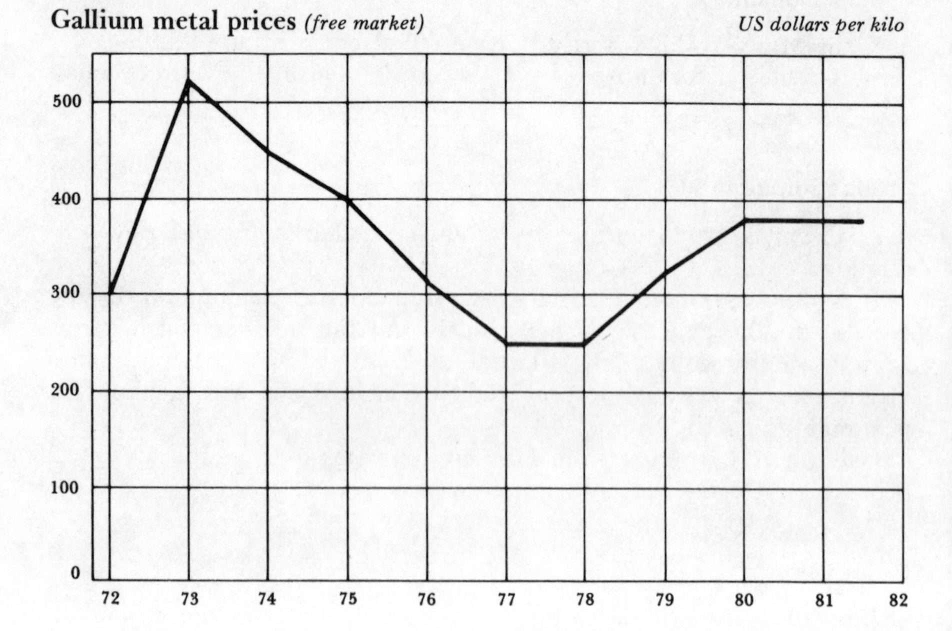

Method of marketing and pricing

There are very few producers of gallium and all are in industrialized countries where the bulk of consumption occurs. Most producers sell at similar prices direct to consumers. The majority are also large producers of aluminium or zinc and have good marketing arrangements in consuming countries.

Prices vary according to grade, but consumers are as interested in technical reliability as much as price, so there is little scope for merchant activity, except in brief periods of shortage or of selling to state buying corporations.

Germanium

Germanium ore production, 1978 *thousand pounds*

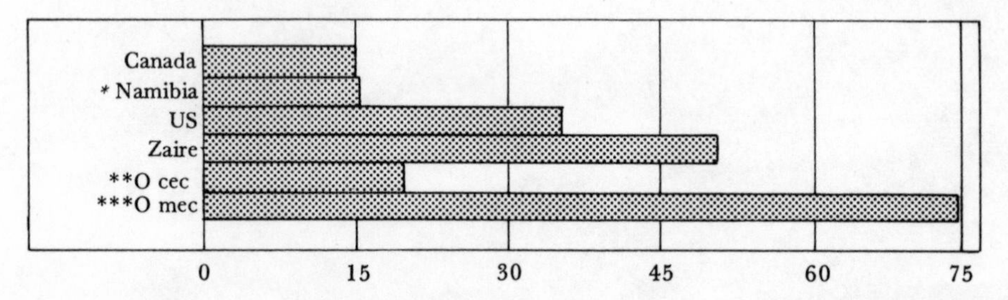

Total world production 195,800 pounds

* S W Africa
** Other central economy countries
***Other market economy countries

Grades available

Almost all germanium is sold in the form of very high purity ingots or in single crystals. Purity is usually not less than 99.999 per cent minimum, but grades are determined normally by electrical resistivity, the most common being 30 ohm cms or 50 ohm cms.

Production method

All germanium is produced as a by-product of other base metal production, usually zinc or copper.

In zinc ore, germanium oxide is volatilized during the refining process and collected electrostatically.

In copper refining the germanium rich residue is collected magnetically as flue dust and the germanium salt is leached out chemically, then smelted to the metal in a hydrogen atmosphere.

A high degree of purity is required for almost all uses of germanium. This is done by standard zone refining techniques in an electric induction ring.

Major uses

Most germanium is used in the electronics industry for various forms of semi-conductors but a quickly growing use for germanium has been found in the manufacture of various night seeing instruments such as binoculars and gun sights. Other small quantities are used for the production of certain solders.

Main market features

Germanium is typical of those metals that have found important uses in the most advanced areas of technology. In the early 1970s it was much in demand for semi-conductors and light emitting diodes. Its use declined in the mid-1970s as new technology replaced much of its use in these areas with other materials. In the last few years, however, more germanium than ever has been required to manufacture nightsighting systems, especially for tanks and aircraft. Consumption for this use alone is responsible for utilizing over half the world's supply and is still growing. Prices have risen dramatically but fresh supplies have been limited because recoverable quantities in zinc and copper ores are so rare and the content of germanium in some power station flue dusts is still too low to be commercially viable. Some new production capacity is under way, however, which may alleviate the shortage somewhat, but it has been predicted that this new production will be completely absorbed by another new use which has been found for germanium, in fibreoptics.

Known world reserves

These are approximately 1,500 tons but this estimate is based on extrapolated production statistics rather than geological data.

Method of marketing and pricing

Because the germanium market is so very small and fickle and marketing so technically orientated, producers need to be well rewarded for their efforts. Prices therefore put germanium into the precious metal class.

Prices vary according to grade and most producers, with perhaps the USSR as the exception, sell at much the same levels. USSR sales are intermittent and, as the price of the metal represents only a small proportion of the value of the component in which it is used, most consumers prefer to rely upon producers who have the best continuity of supply and sound technical reliability.

Merchants are increasingly involved in the trade of germanium, especially between state-controlled producers and consumers.

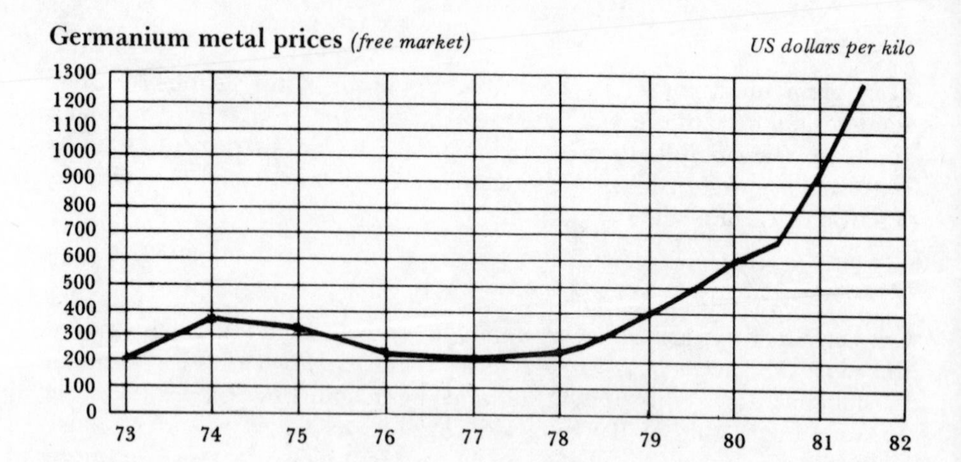
Germanium metal prices (free market)
US dollars per kilo
1300
1200
1100
1000
900
800
700
600
500
400
300
200
100
0
73
74
75
76
77
78
79
80
81
82

Indium

Indium metal production, 1980

hundred thousand troy ounces

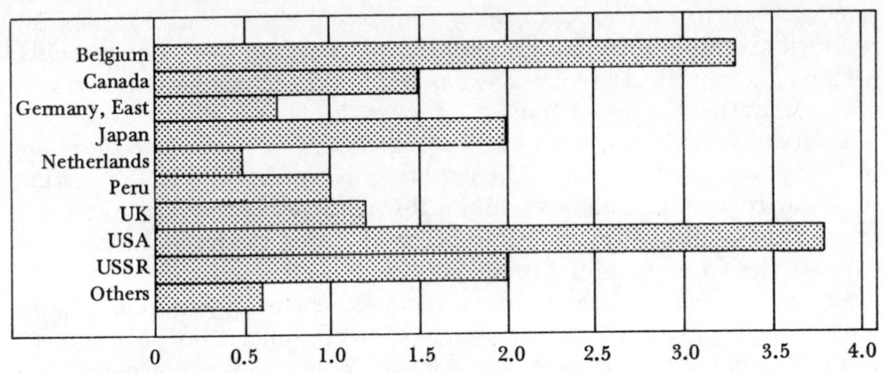

Total world production 1,600,000 troy ounces

Grades available

Indium is normally traded in the form of bars or ingots with a minimum purity of 99.99 per cent but some producers still offer a 99.97 per cent grade which is traded at a discount to the normal price.

Production method

All indium is produced commercially as a by-product of zinc and lead production, but only some zinc and lead ores contain indium in recoverable quantities.

The indium can be recovered by a number of methods, the most common being distillation, but it is also recoverable by leaching and electrolysis.

Major uses

Indium has a wide range of uses: as a thin alloy layer in high performance bearings, mainly for sports cars; in the electronics industry, as a component of transistors and in computer equipment; in low melting point alloys and in solders.

Some of these low melting point alloys are used for glass lens grinding and polishing, and as plugs for fire sprinkler systems.

Indium is also used in catalysts and in the purification system for the manufacture of certain chemical products, particularly man-made fibres. The metal also has many minor uses in research and in laboratory techniques. Small quantities are also used in nuclear reactors.

Main market features

Practically all the uses for indium are either technologically sophisticated or obscure. The electronics industry is notorious for rapidly revolutionizing systems and completely changing its requirements for raw materials. Indium has suffered setbacks in demand as a result of this over the last few years.

Demand from other areas has fluctuated also but, because there are so many uses for this metal, the net effect has been to keep the market surprisingly steady. Like germanium the increase in price is attributable to increased demand, this time from the electronics industry, and because the zinc ore mined these days contains less indium. The mining of zinc ores which contain useful and expensive by-products is an obvious way for mines to increase their profitability. But such ores are comparatively rare and over the years the content of these metals has diminished, just as the cost of extracting them has increased. Where these by-product metals are used in electronics and for other sophisticated uses, the price of the by-product is bound to increase, especially when the production of the host metal, used mainly in less healthy sectors of the economy, is being curtailed due to lack of demand.

There has been an element of 'speculative' buying interest in indium in recent months, however, which has meant that the indium price 'has drifted' as speculators take their profits but high prices must be expected for the foreseeable future.

Method of marketing and pricing

Canada is the largest producer of indium and, together with the producers in the US, maintains a strong but responsible influence on the market. As with many other metals, there are very few large consumers of indium and a multitude of small consumers. The large consumers almost invariably sign their supply contracts with one or other of the large North American producers. This is understandable, as a continuity of supply as well as technical reliability are almost as important as the price.

The price paid by these consumers provides a pricing basis for other producers when selling to smaller consumers but a discounting or premium system exists depending on the state of the market.

There is a small but fairly active free market conducted by the few merchants who specialize in this type of rare metal. Their main source of material is the USSR which conducts a sporadic sales policy. Supplies are also available from Japanese and South American producers who prefer to sell to international merchants rather than to set up their own marketing operation which would be disproportionately expensive owing to the large number of smaller consumers who need to be serviced.

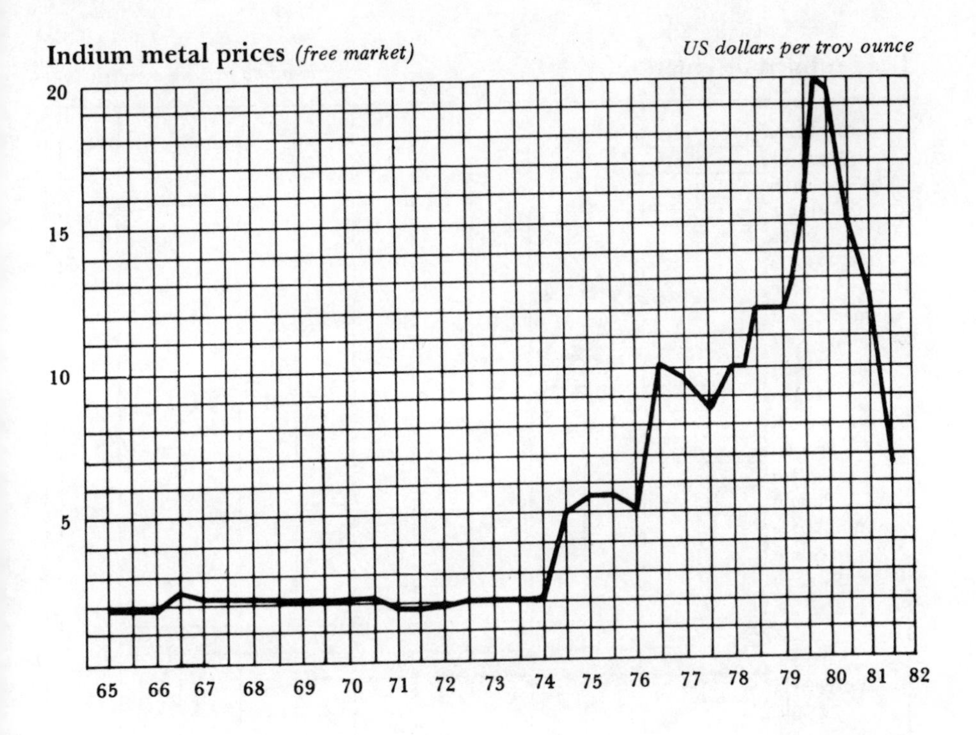

Lead

Lead refined production, 1980 *hundred thousand metric tons*

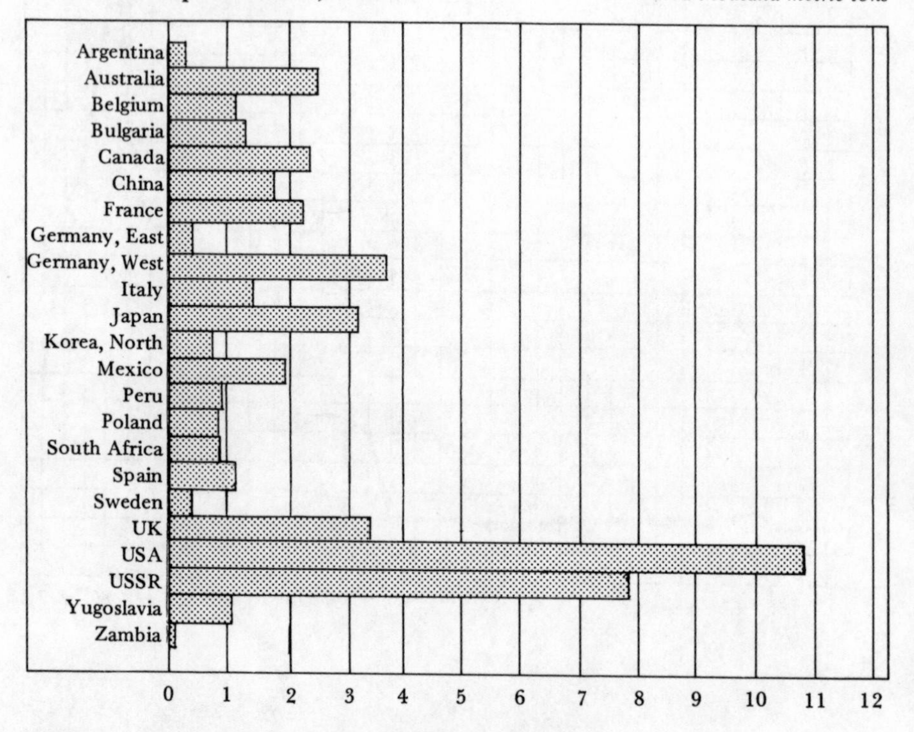

Total world production 4,011,000 metric tons

Grades available

The most commonly traded grade of refined pig lead is 99.97 per cent minimum. This is the standard minimum grade which can be traded on the London Metal Exchange. There is a large trade in secondary lead (remelted from scrap) which normally has a purity of 99 per cent minimum.

Production method

Lead is found in several minerals, but its most common ore is galena (lead sulphide). Commercially viable lead ores can also be associated with certain zinc bearing minerals. Preparation for smelting varies with the grade of ore. Some high grade material may not need treatment, but low grade ores are concentrated by flotation. The high grade ore concentrate is roasted in air to remove the sulphur, then smelted in a blast furnace or open hearth furnace with coke to reduce the oxide to lead bullion with a purity of about 97 per cent. The bullion is further refined by removing other metal impurities by skimming, after the addition of various reactive chemicals.

Major uses

There is a doubtful future for the use of lead in three important areas. Its most important use in car lead/acid batteries may be short-lived as most authorities believe that lighter, more efficient batteries will

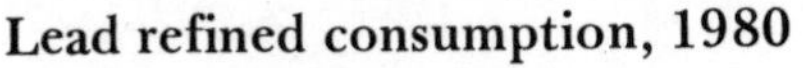

Lead refined consumption, 1980 *hundred thousand metric tons*

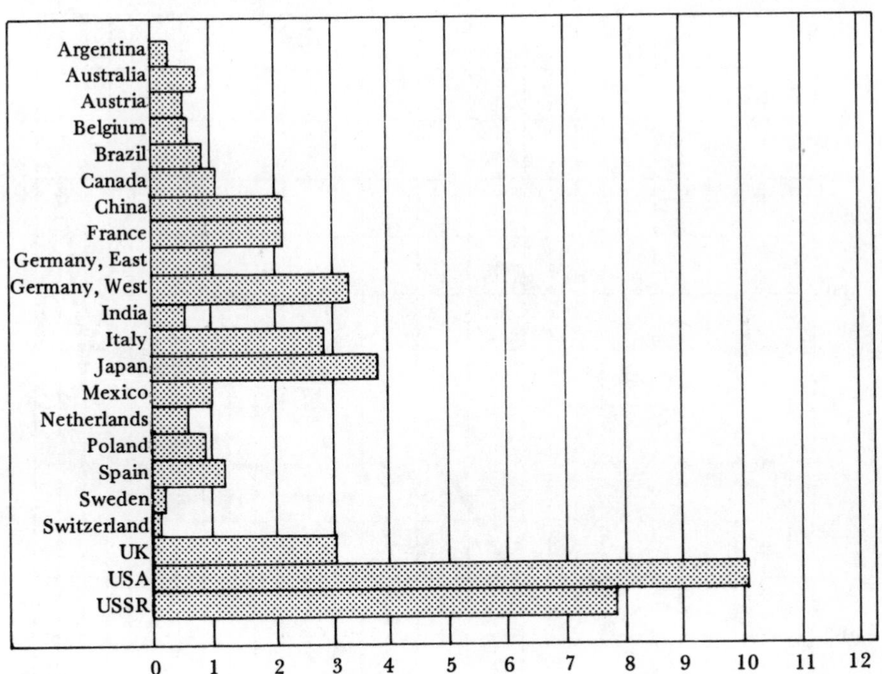

Total world consumption 3,806,000 metric tons

replace the lead battery within a decade or so. Its second major use as an additive to petrol is coming under greater scrutiny due to pollution problems. A minor, but still important use in type metal for printing is fast becoming obsolete.

Lead uses for making solders, ammunition and anti-corrosive sheaths for electric cables, however, are still strong markets. Other important minor uses for lead are in anti-radiation shields and as salts in the chemical industry.

Main market features

Most lead is consumed in the countries in which it is produced. Although it has some interesting uses it has the disadvantages of being soft, heavy and toxic. Its softness makes it unsuitable for structural purposes and in batteries it must be alloyed with antimony or calcium. Its density has prompted research into finding new, lighter car batteries which will save fuel and its toxicity has drastically reduced its uses as an

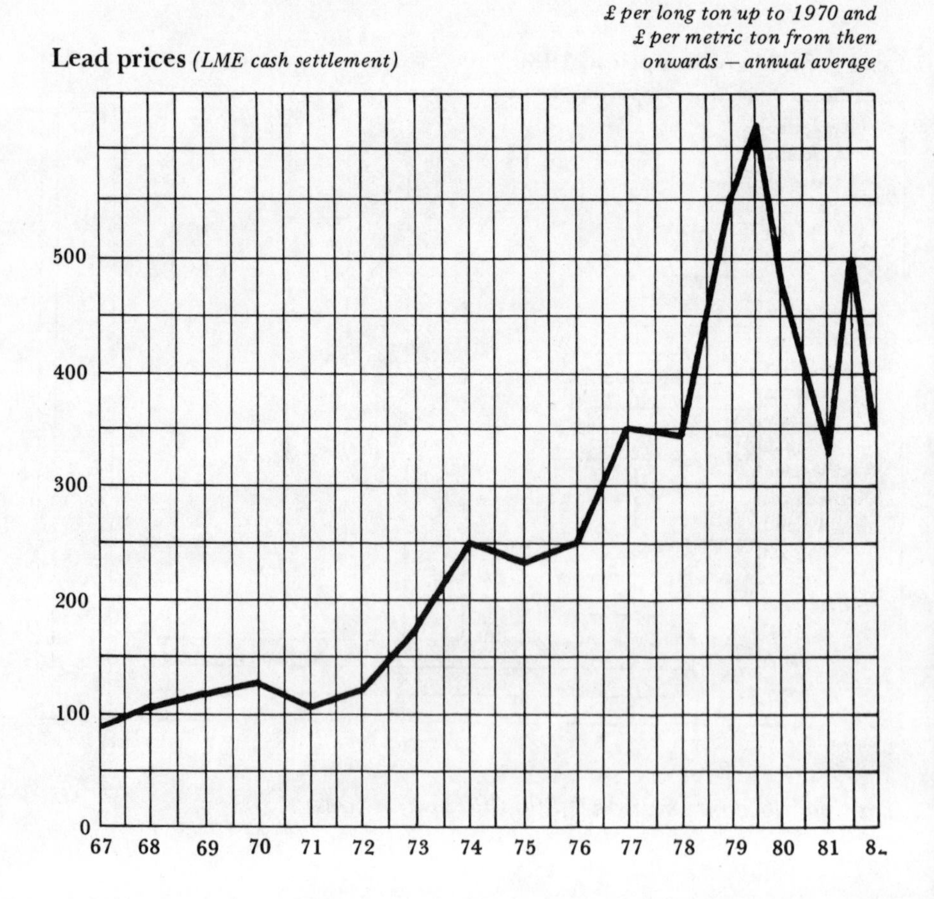

'anti-knocking' agent in petrol, in water pipes and many other uses.

Lead is an excellent radiation shield and finds many uses in the field of nuclear energy and research and where isotopes and X-rays are used. This has provided an area of increased consumption, but, with so many major uses of lead threatened with replacement by advancement of technology, the future for the metal is not bright.

Scrap lead derived from used car batteries is an important supply source of the metal.

Method of marketing and pricing

The producer price system is still used for much of the world's trade in lead. But these prices only usually prevail in the domestic market of the particular producers and move regularly in line with the free market as reflected in the prices traded on the London Metal Exchange.

A large world trade is conducted in lead concentrates, most of which is conducted on an LME price basis. Heavy buying of the metal on the LME by the USSR has raised prices considerably from time to time over recent years.

Lithium

Lithium ore production, 1977 *thousand metric tons*

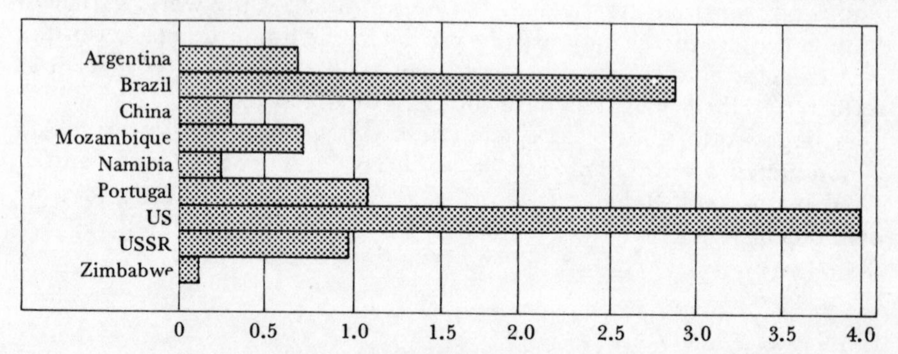

Total world production 10,770 metric tons

Grades available

Pure lithium metal is seldom used or traded. It reacts violently with water so must be kept in strictly dry conditions.

Most applications are for lithium compounds, the most common being lithium carbonate. Other useful compounds are hydroxide, chloride and fluoride.

Production method

Lithium's major ores are petalite, spodumene and lepidolite. It is also found in recoverable quantities in certain brines. The compounds are recovered from the ore by chemical means, either by an acid or alkaline treatment.

These are recovered from brines by a leaching and flotation method after concentration.

Major uses

Lithium is the lightest of all metals, but its uses are limited because of its reaction to water. Small quantities of the metal are used for making light weight alloys in combination with magnesium and aluminium. It is also used as a scavenger and degasifier in the production of steel and as a deoxidizer in melting copper and copper alloys.

The main uses for its compounds are as follows:-

a) *Lithium carbonate* This is used for ceramics, particularly enamels. A comparatively recent but major use is in aluminium smelting where it is added to the molten alumina. Here it has the effect of reducing the temperature at which electrolysis takes place and improves the yield of aluminium production.

b) *Lithium hydroxide* The major use for this compound is in the production of greases.

Lithium compounds are also used in the manufacture of bleaches, disinfectants, synthetic rubber, in welding, brazing and in the glass industries.

Main market features

The only major producers of lithium and lithium compounds are the US and USSR and to a certain extent West Germany. The two major US producers effectively control prices for the Western world but they have to compete with Russian exports outside the US.

There are signs, however, that the USSR is consuming more of its own production lately, thus reducing competition for Western producers.

The US has huge reserves of lithium ore and lithium containing brines, and is not dependent on imports of raw materials.

Reliance on imports of feedstock would limit future production outside lithium ore producing countries.

Known world reserves

6,036,000 tons, of which approximately 4,000,000 are estimated to be in the US.

Method of marketing and pricing

Merchant activity is small and confined to the marketing of USSR material, but the Russians are only intermittent sellers on the world market, so most consumers only buy Russian material to 'top up' supplies.

There is a small amount of competition between Western producers but their prices are usually very similar.

Lithium metal prices *(producer price)* *£ per kilo*

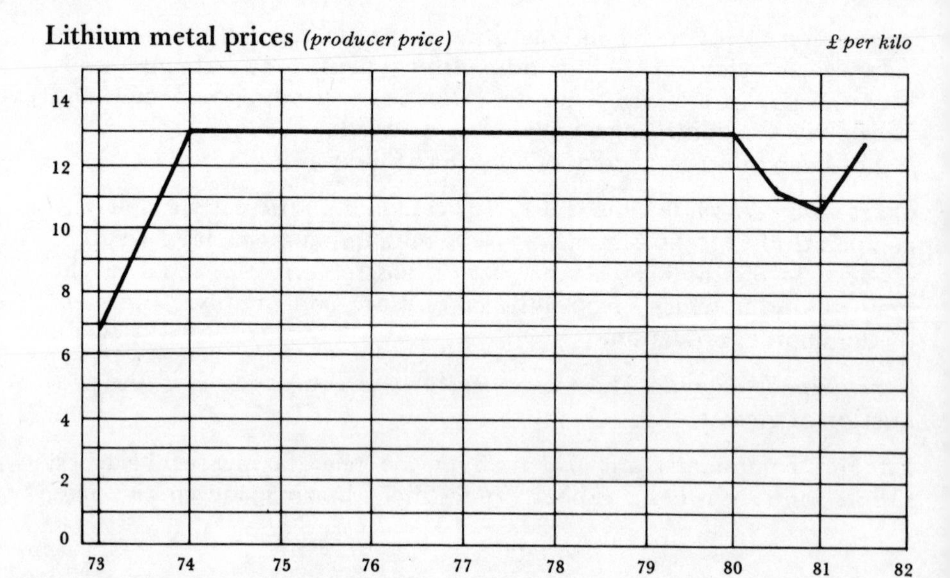

Magnesium

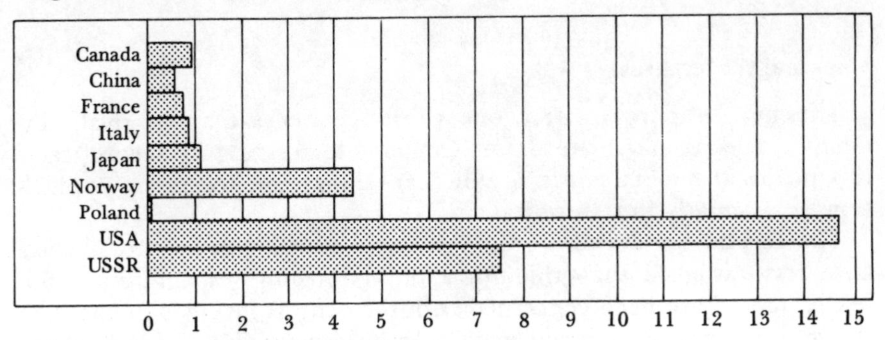

Total world production 313,000 metric tons

Grades available

Magnesium is traded in the form of ingots usually between 5-20kg each with the minimum purity of 99.8 per cent or 99.9 per cent.

Production method

Magnesium is found naturally in the ores magnesite and dolomite, the latter being its principal source in the earlier days of its commercial exploitation. It is now more commonly extracted from brines in salt lakes or seawater. This innovation is likely to represent almost all future production.

Extraction from the ore The ore is first crushed, then thermically reduced with ferro-silicon at high temperatures in retorts. Magnesium comes over as a vapour which is then condensed.

Extraction from brines (magnesium chloride) The chloride is melted in a steel electrolytic cell into which graphite anodes are suspended. The magnesium is released by the passage of a direct electrical current and rises to the surface of the melt where it is tapped into ingot moulds.

Major uses

Magnesium is a metal similar to aluminium, in that it is light and its alloys are strong. It also has many of the same applications.

Magnesium is used in alloy form to make castings, particularly for the aircraft and automobile industries. It is also used in small proportions as an alloy addition in certain aluminium alloys: one of these alloys is used to make the ring-pull type can tops.

Another very important use is in the refining of other metals, notably titanium, where it is used to reduce titanium tetrachloride to the metal.

Ferro-magnesium, usually in the form of ferro-silicon-magnesium, is used in the manufacture of certain cast irons.

Main market features

Magnesium is approximately one third less dense than its rival, aluminium, and its alloys are almost comparable in strength. Theoretically this makes the metal more useful than aluminium for many applications, particularly in transportation.

The technology for the production and working of magnesium has been less advanced than that for aluminium, however, and many potential users have been concerned with the dangers involved in handling the molten or finely divided metal – both can oxidize violently in air. New fluxes and chemical covers can, however, minimize these dangers to a level comparable with those encountered in using aluminium. This should encourage more applications.

World production of magnesium is less than one twentieth of that of aluminium, which automatically increases its unit cost. It is this higher cost which has been the main reason for its comparative lack of development. However, consumption has been growing at a comparable percentage rate with that of aluminium. Magnesium manufacturers have long predicted a future boom in magnesium production once the production grows large enough to reduce unit cost.

Unfortunately, there are very few magnesium producers, partly for the reasons outlined above and partly because production should ideally be sited in an area of low electricity costs. With the advent of production from seawater, however, more sites may be considered suitable.

Prices cannot be expected to rise substantially unless the aluminium price rises disproportionately or unless titanium production is drastically increased.

Known world reserves

Reserves of magnesium are limited only by man's ability to extract it from seawater.

Method of marketing and pricing

With the exception of the USSR, production is confined to industrialized Western countries, which have so far managed to retain a tight grip on the price, particularly as it applies in the country of production.

The rising price of fuel has encouraged motor car manufacturers to reduce the weight of vehicles. As magnesium is a very suitable metal to produce the kind of castings used in motor cars, such as gearbox casings, wheels and even engine blocks, and is an extremely light metal, it is now becoming increasingly popular in the motor car industry. In addition, the high price of titanium is stimulating higher titanium production, which requires magnesium as a fuel. Magnesium consumption is therefore growing. A high price for aluminium will also encourage producers to increase the magnesium price.

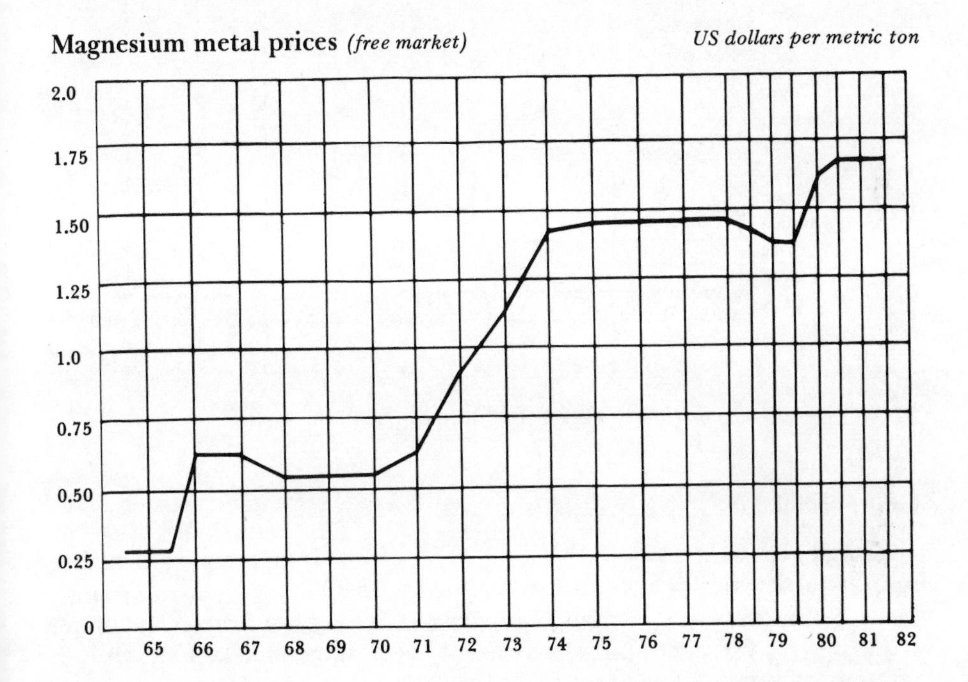

Magnesium metal prices *(free market)*

Manganese

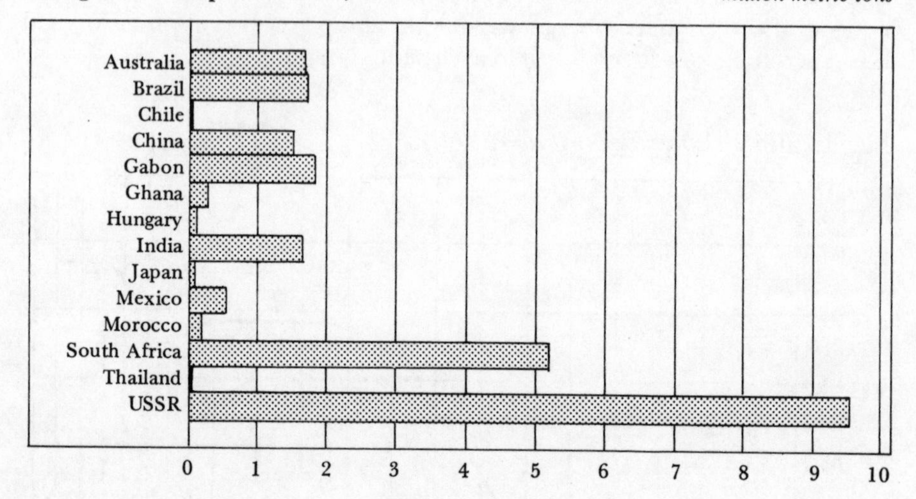

Total world production 24,455,397 metric tons

Grades available

Pure manganese, usually in the form of an electrolytic lump, is available with a purity of 99.8 per cent minimum or 99.9 per cent minimum. More common forms of manganese on the market are ferro-manganese, with up to 80 per cent Mn content, and silico-manganese, with a content of up to 20 per cent Si, balance Mn. Prices for ferro-manganese vary with the Mn content and carbon contamination.

Production method

Manganese is found naturally in the form of a number of different ores. The ores are concentrated by normal methods which vary according to the type of ore.

Ferro-manganese is produced by smelting the concentrate in either a blast furnace or an electric arc furnace.

Major uses

About three quarters of all manganese produced is consumed in the form of ferro-manganese used in the manufacture of manganese-containing irons and steels. These are in turn used in a wide variety of applications covering the entire spectrum of industry from construction steels used for bridges, to the steel used for home appliances.

Small proportions of pure manganese are used as an addition to certain aluminium alloys.

Manganese compounds are used in the manufacture of dry cell batteries, in the manufacture of some dyes and bleaches and also to produce several other important chemicals.

Main market features

Although the USSR and South Africa together produce about half the world's manganese (a feature shared with several other metals) there are a number of other important producing countries. Ferro-manganese is produced in an even larger number of countries. It is very difficult for any one producer to control prices, especially as manganese is such a large market.

There is no strict relationship between the price of the ore and ferro-manganese because of the varying cost of production but, because demand for manganese accurately reflects the world's general economic climate, the prices of both have moved historically in unison.

The market in individual consuming countries can be confused because of changing import tariffs.

Known world reserves

Reserves of manganese contained in ore approximately 728 million metric tons. This does not include manganese which may be found on the ocean floor in the form of nodules.

Method of marketing and pricing

Most ferro-manganese is consumed by large companies (iron and steel producers). They usually buy their supplies on long-term contracts directly from producers or producers' agents, or from a merchant, who has in turn bought long-term supplies from the producer. The reason for this style of business is understandable when one considers that ferro-manganese is a very cheap commodity traded in comparatively large tonnages. In order to make prices competitive, material must be shipped directly from producers to the consumers' works. Profit margins per metric ton are generally so small that if material has to be stored awaiting a buyer, or shipped to a port further away than the one originally intended, the extra costs would not be met by profit margins.

The largest producers sell at a producer price but smaller producers and merchants trade on a free market basis at discounts or premiums to the producer price, depending on the state of the market.

Each year the major producers, led by the South Africans, confront their consumers with higher prices, mainly for manganese ore but also for other manganese products. Their success in achieving higher prices depends very much on the ruling market conditions.

There is a small, but active, market in ferro-manganese between merchants and smaller consumers such as cast-iron foundries.

The manganese metal market is not particularly interesting. The metal is consumed mainly by secondary aluminium alloy makers (those utilizing aluminium scrap in their feedstock). There are few large companies of this type, so their intake of manganese is usually only a few metric tons per month. There are merchants who specialize in this type of business, but again they have to compete with producers who sell to the consumers either directly or through agents.

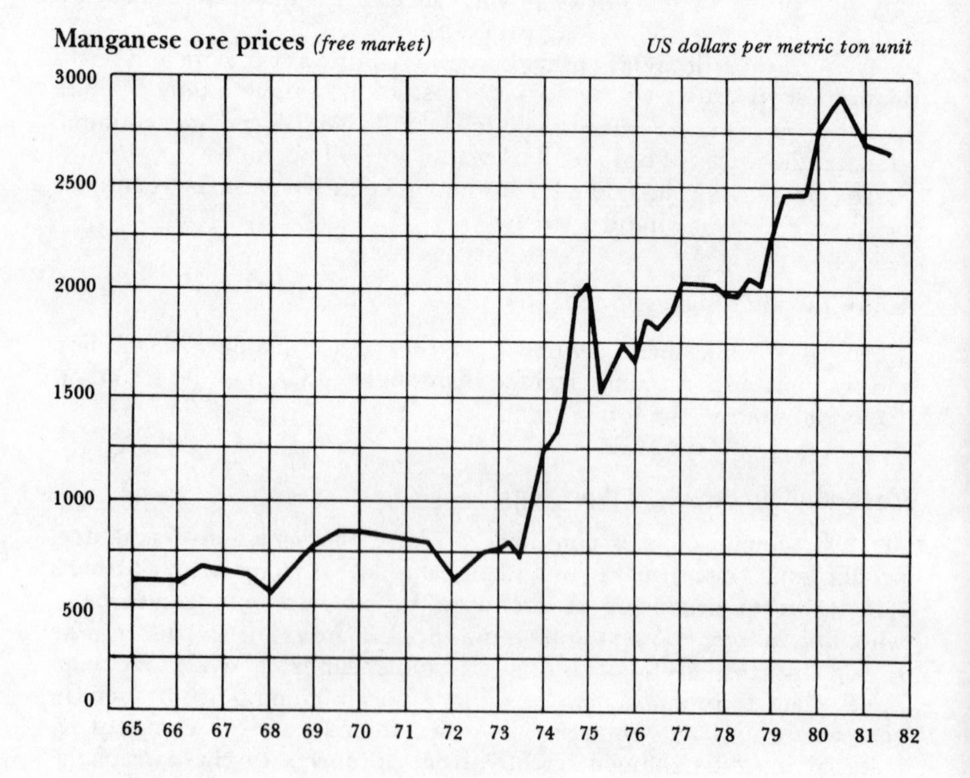

Mercury

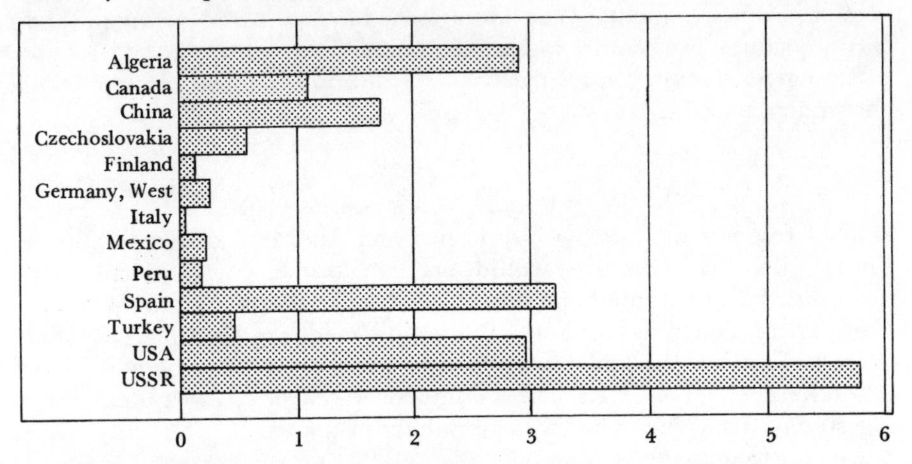

Total world production 181,105 flasks

Grades available

Mercury is sold by the flask, the steel bottle containing 76 lbs of the liquid metal. The standard purity is 99.99 per cent minimum.

Production method

Mercury is produced from its naturally occurring sulphide ore, cinnabar.

The ore is crushed but no attempt is made at further concentration. It is then roasted in retorts and the mercury metal is vapourized and collected. It is a very simple process. In some areas of the world mines produce only two or three flasks per week and are worked by only one or two miners who both excavate and refine the ore.

Major uses

Mercury is an extremely toxic metal. All its uses have come under increasing scrutiny because of its possible effect on the environment.

The best known use for mercury is in thermometers, barometers and pressure gauges, but this accounts for only about 10 per cent of its consumption.

It has three major uses. The first is in the manufacture of chlorine from brine, but other methods can be used to produce chlorine and these are gaining favour for environmental reasons. The second is in batteries, but other types of batteries provide adequate alternatives. The third major use is in agriculture as pesticides and fungicides, but here too environmental considerations have reduced demand and encouraged the use of alternative substances.

Strangely, perhaps, mercury has several important uses in medicine and in dental alloys.

Main market features

Known reserves of mercury are rather small and led to speculation in the 1960's that mercury would become scarce. Such speculation, however, did not predict the recent international concern about pollution. These concerns were reinforced by evidence that mercury has caused many deaths and much disease, the more serious of which are a result of chemical salts being dumped at sea or in rivers and then digested by fish. The mercury accumulates in the fish's body and is thus passed on to humans.

Mercury is the only metal which is in liquid form at normal temperatures and will always find a use for that reason, but where alternative materials can be used they will be used. And these alternative materials probably account for over half the world's current consumption. This notion combined with very low prices has caused most producers, particularly in Spain and Italy, drastically to curtail production and sales. This policy has begun to have some success, especially since the Chinese seem unofficially to have adopted a strategy similar to the main Western producers.

Known world reserves

Commercially viable ore contains approximately 3,200,000 flasks.

Method of marketing and pricing

There are no effective international producer prices for mercury. All producers have to sell at a free market price, but many large orders for the metal are purchased on a tender basis, where merchants and producers alike submit offers and the cheapest offer wins the contract. In the depressed circumstances of the mercury market consumers have had the upper hand.

In the 1960s there was a very active and important market in mercury which attracted many large trading houses. With the combination of low prices, low turnover and the absence of price movements, there are now few merchants specializing in this trade.

Scrap recovery

There is a fair trade in scrap mercury, usually carried out on a local basis. This material can be cleaned by filtration or by distillation.

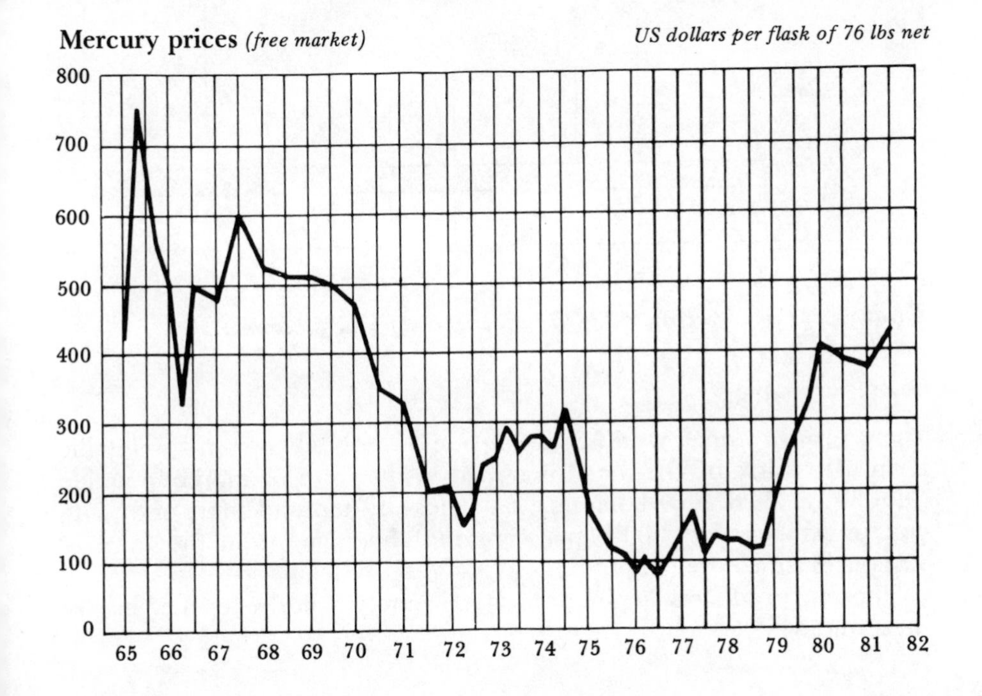

Molybdenum

Molybdenum ore production, 1979 *ten thousand metric tons MoS_2 content*

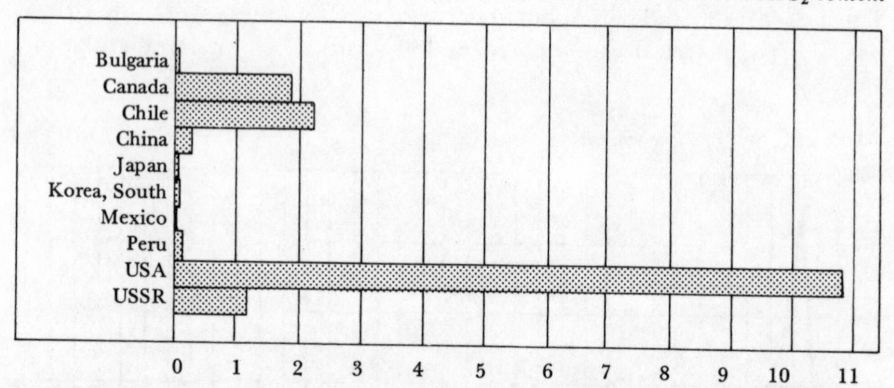

Total world production 172,000 metric tons MoS_2 content

Grades available

There is only a small world consumption of molybdenum in its metallic form. The bulk of the metal is traded in the form of molybdic oxide (Mo03) or ferro-molybdenum, the alloy of molybdenum and iron, containing between 60-80 per cent molybdenum or in the form of its concentrated ore (sulphide).

The few producers of the pure metal supply it in the form of ingots, sheet and wire.

Production method

The naturally occurring ore, molybdenite, a molybdenum sulphide, is concentrated by flotation. The oxide is produced by heating the sulphide in air. The oxide can then be purified by sublimation. The material is produced by the reduction of the pure oxide in the presence of hydrogen at high temperatures. Between one quarter and one third of the world's production of molybdenum is produced as a by-product of copper production when the copper ore contains a high proportion of molybdenum.

Major uses

The pure metal has a small but increasing application in highly technical industries such as nuclear energy, aerospace and electronics.

Its main application by far is as an alloy addition to steel. Its effect on steel in additions of about 1 per cent is much like that of chromium and nickel: it increases strength and toughness especially at high temperatures. Such steels find applications in the production of mining equipment and machine components and have a growing use in oil pipelines.

Major market features

The US is much the largest producer of molybdenum and it is produced there by only a few companies. These companies are therefore in a powerful position to control the market price. They publish a producer price for their major products which is adhered to but they ration material to their customers in times of shortage. Supply is somewhat affected by fluctuations in copper production but, strangely, not much affected by the steel market variations due to the special applications of molybdenum steels, the uses of which may be somewhat out of phase with general economic conditions. The development of the oil industry is particuarly relevant to demand due to its application in pipelines and refineries.

Demand for molybdenum from every sector increased tremendously in the last year or two. It completely outstripped supply and caused a major shortage and an attendant remarkable price rise. High prices have, however, stimulated increased production ventures, which have brought supply and demand into balance. Speculation in molybdenum exacerbated the shortage in 1979 and the sales of these stocks has helped to bring the price down to more reasonable levels. One cannot expect much of a further fall in price, however, as production costs increase and demand remains high.

Known world reserves

Approximately 4,900,000 metric tons of which 3 million metric tons are estimated to be in the United States.

Method of marketing and pricing

Molybdenum prices in the form of sulphide, oxide and ferro-molybdenum are quoted as a price per pound or per kilogram of molybdenum content.

Free market trade is somewhat intermittent, drying to a trickle in times of supply/demand balance but flaring up into hectic trading in times of shortage. This trade is conducted by comparatively few international trading houses which usually specialize in all the refractory metals, that is, vanadium, columbium, titanium and tungsten.

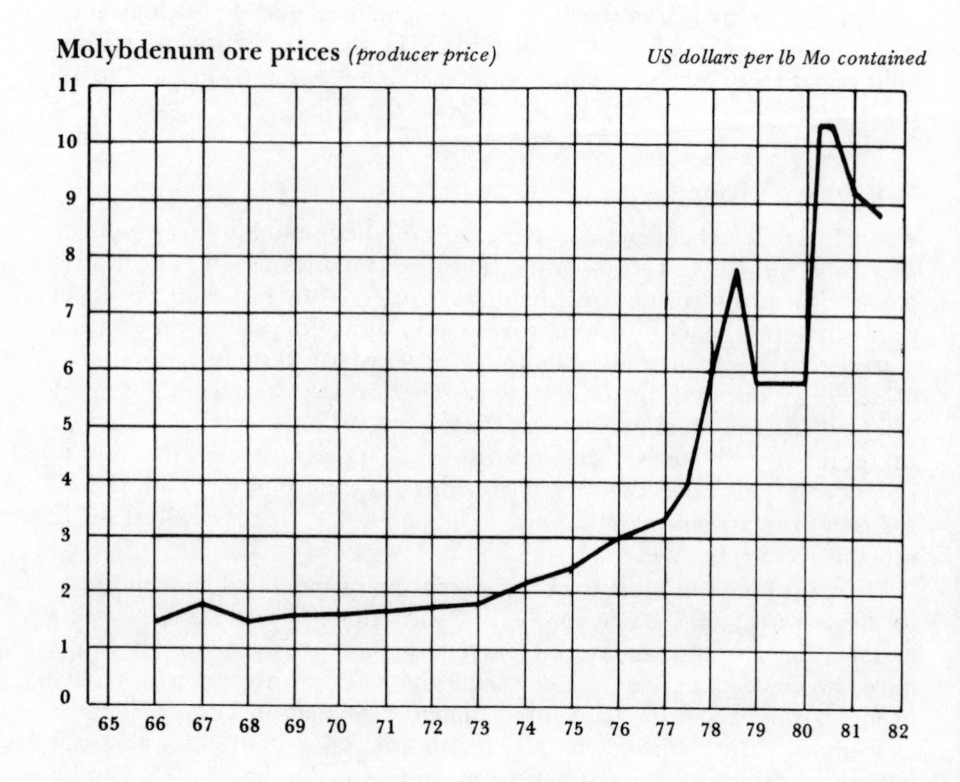

Nickel

Nickel metal production, 1980

ten thousand metric tons

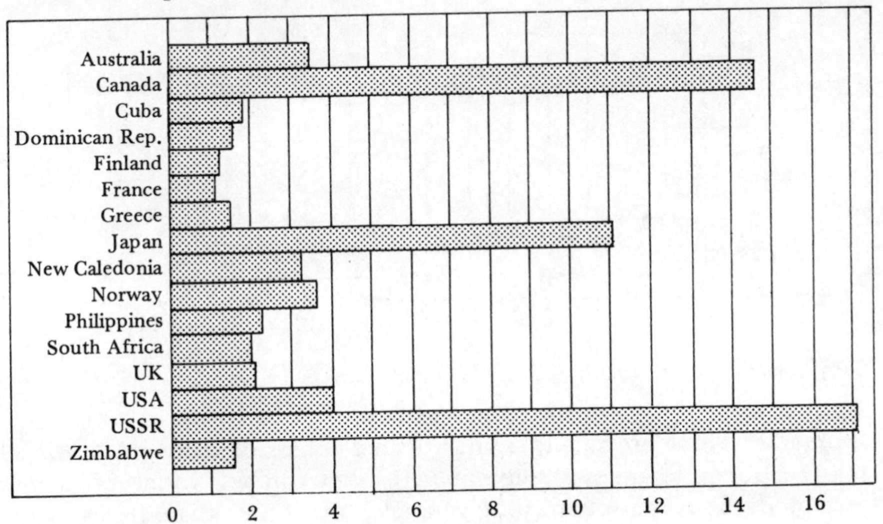

Total world production 751,100 metric tons

Grades available

Nickel comes in many shapes, sizes and purities to suit the various uses to which it is put.

Cathodes These are produced by electrolytic means and are either full size (un-cut) up to 750mm x 750mm or cut into smaller squares for convenience. These squares range from 1in x 1in up to 10ins x 10ins. The full size cathodes are cut into squares using a power shear, so they are only roughly square. The smaller squares are always more expensive than the larger ones as more cuts have to be made. As a very rough guide, 10ins squares are about two US cents per pound more than un-cut cathodes, but 1in x 1in squares may be 10 US cents per pound or more above the price of 10in x 10in squares. The thickness of cathodes varies with the producer, but generally is about 1cm.

The smaller squares may be used for electroplating.

Minimum purity is usually 99.8 per cent with a maximum cobalt content of 0.1 per cent.

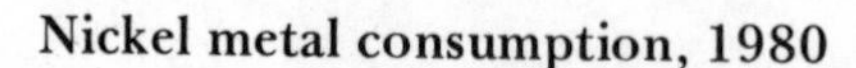

Nickel metal consumption, 1980

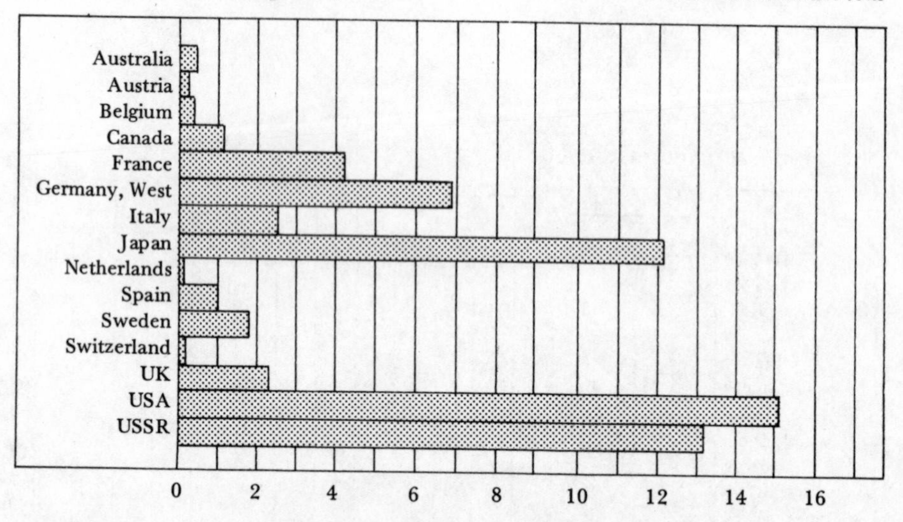

Total world consumption 704,700 metric tons

Briquettes These are small, regular shaped lumps of compressed nickel matte (spongy granules of pure nickel). They can be a variety of shapes – tablet, round or square, but generally no bigger than 2½ins in the largest dimension.

Granules These are solid pure nickel in granular form.

Pellets and Shot These are made by the largest producers.

Ferro-nickel An alloy of nickel and iron usually containing about 30 per cent nickel.

Nickel oxide This is known sometimes as sinter, usually in briquetted form.

These are the main types of nickel traded. There are premiums and discounts depending on the type of material: cathodes and pure pellets are the most expensive and ferro-nickel the cheapest. All prices move up and down in unison. To complicate the situation further, some origins of metal may be more expensive than others. The best guide to price movements is the market for Russian 10in x 10in cathodes which are probably the most commonly traded on the free market.

Production method

Almost all nickel is mined as nickel ore rather than as a by-product of other metal ores. The ore is either an oxide or a sulphide of nickel. The sulphide ores are crushed and concentrated by flotation but with laterite (oxide ore) no concentration method has been found effective.

Where nickel is found in conjunction with copper, the nickel may be separated by a combination of magnetic and flotation techniques and by recrystallization. Sulphides are converted to oxides by roasting in air and the oxides reduced to the metal in a reverberatory furnace. Where nickel ore contains cobalt or other secondary metals separation is by chemical means.

Major uses

Nickel has a vast number of uses throughout industry. The most commonly known use must be in coins when alloyed with copper, and in stainless steel when alloyed with iron and chromium. It is used in metallic and salt form with and without other metals and salts, as a catalyst in a large range of inorganic chemical reactions. Many types of steel alloys contain nickel especially when resistance to corrosion is required. It is used for plating, especially in the automobile industry and for cutlery. Some types of batteries use nickel. It forms part of many high strength alloys used in the aircraft industry. Stainless steel can be made with several other metals but nickel is by far the most competitive in cost.

Main market features

Historically nickel has been produced by a limited number of powerful groups led by International Nickel of Canada, who at one stage had over 80 per cent of the western world market. However, the last 10 years have seen dramatic changes in the structure of the industry with the entry of many newcomers and International Nickel's share has been cut down to below 35 per cent. The turnround started in 1969 when a long strike at the Canadian mines of Inco, and Falconbridge, resulted in a severe shortage of supplies. This forced consumers to look for alternative sources, or substitutes for nickel, and attracted new producers into what then became a highly competitive market. Canadian sulphide deposits remain the main source of nickel at present, but the bulk of future reserves are in laterite deposits.

Method of marketing and pricing

The London Metal Exchange commenced nickel trading in 1979. This move coincided with another short strike in Canada's nickel mines.

The combination of these events, stimulated by speculative interest through the LME, had the effect of increasing the price dramatically, even though there was no physical shortage of the metal throughout this period.

The LME is now by far the biggest influence on the nickel price, although most of the larger consumers still buy their requirements directly from the producers.

The price of nickel has remained comparatively steady over the last few years, which means, of course, that it has failed to keep up with the rate of inflation. This has been due partly to increased competition from smaller producers and partly to the fall in demand during the current recession.

Producers are likely to reduce output to avoid building up stocks which must be financed in a period of high interest rates in order to maintain the price at its present level.

Known world reserves

Approximately 90 million metric tons.

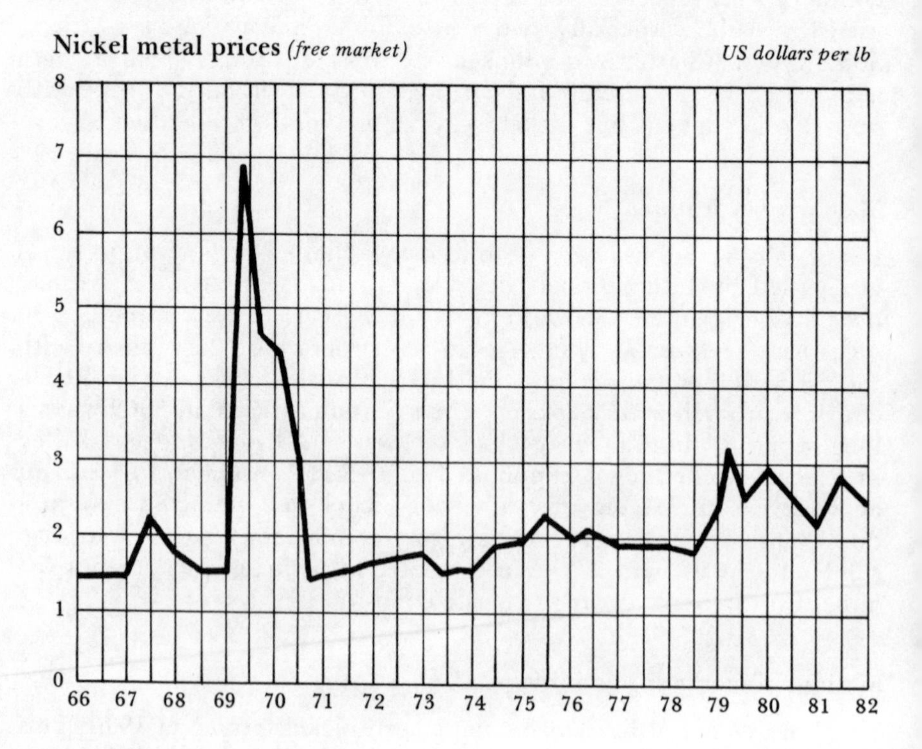

Rhenium

Rhenium metal production, 1978

thousand kilos

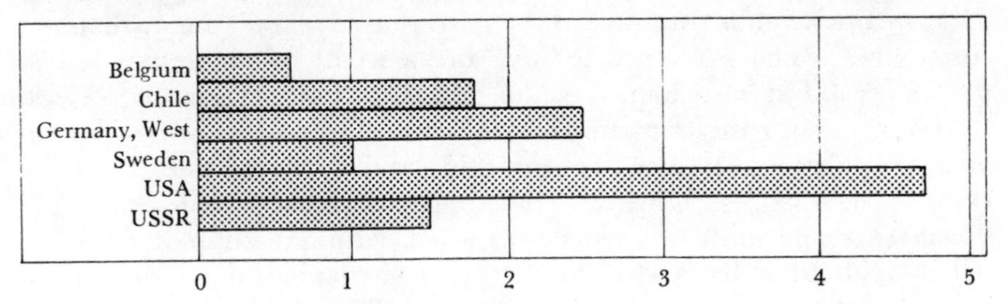

Total world production 12,100 kilos

Grades available

The major rhenium producers offer a variety of metallic forms and compounds but most international trade is conducted in metallic powder with a purity of 99.99 per cent minimum and in ammonium perrhenate NH_4ReO_4 containing 69 per cent rhenium.

Production method

Rhenium was isolated comparatively recently, in 1925, and has only been used commercially since the early 1950s. It is produced commercially only as a by-product of molybdenum production, where it occurs in copper-molybdenum porphyry ore bodies. It is recovered from the flue ducts and gases produced during the roasting of the molybdenum concentrates.

Major uses

The most important uses for rhenium are as catalysts used to reform petroleum, in the production of high octane petrol and in the production of benzene, toluene, xylenes and other organic compounds.
Its other important uses are in lamp filaments, thermocouples and electrical contacts.

Main market features

Although rhenium is not a platinum group metal, it has very similar physical and chemical properties to those of platinum. Its uses are similar but, unlike platinum, which is found in recoverable quantities only in very few areas, rhenium is found in most copper-molybdenum ores.

Some 30 tons of rhenium contained in ore are mined every year but only about one third of this production is extracted, either because the molybdenum producer has no extraction plant or because low rhenium prices in the past have made extraction unprofitable. However, during 1979 prices trebled. This was because rhenium can be used as a substitute for platinum. In practice rhenium is used in a platinum/rhenium compound which not only reduces the amount of platinum used in the catalyst but also improves the efficiency of the catalytic effect. Platinum prices rose dramatically as a result of a combination of investment speculation, increased consumption and restricted sales from the USSR. High prices, however, encouraged increased production and speculators sold most of their stocks when platinum group metal prices fell in mid-1980. By mid-1981 the price of rhenium had collapsed to about a quarter of its price in the winter of 1979-80.

Rhenium could be used in a much wider range of catalytic reactions and consumers would have the advantage of a cheap replacement for platinum group metals and would not have to rely on South Africa as a source, but, while production remains low compared to platinum, catalyst consumers are reluctant to take advantage of these factors. Any future shortage of platinum group metals could help to establish rhenium as a more important metal.

Known world reserves

Approximately 3,500 tons, of which 1,300 tons are in the USA and 1,200 tons in Chile.

Method of marketing and pricing

Until the increased demand for rhenium occurred in 1979 there was almost no merchant activity in the rhenium market. The shortage of metals has encouraged a fairly active market, however, conducted by specialist metal merchants who obtain most of their supplies from the USSR and the smaller molybdenum refineries. Small quantities are recovered from the treatment of scrap rhenium.

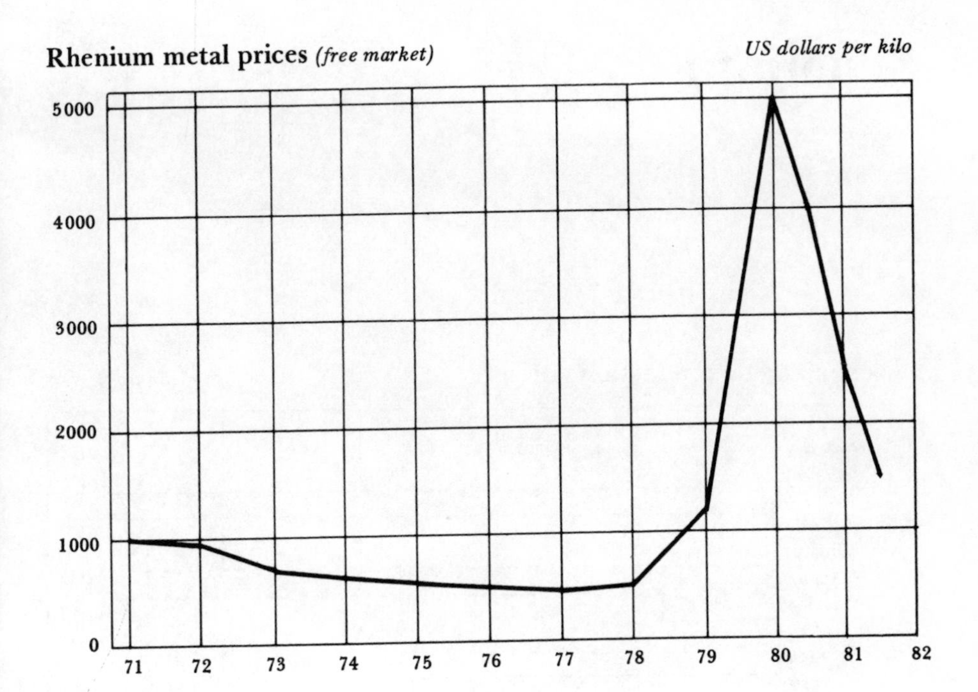
Rhenium metal prices (free market)
US dollars per kilo
5000
4000
3000
2000
1000
0
71
72
73
74
75
76
77
78
79
80
81
82

Selenium

Selenium metal production, 1979* *hundred thousand kilos*

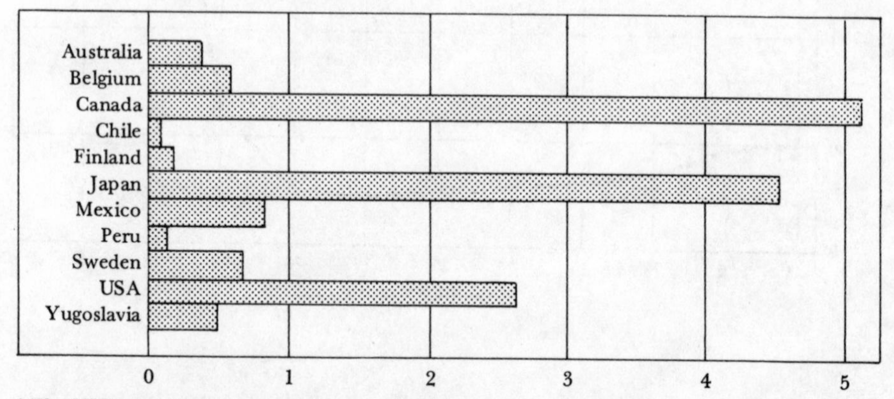

* The USSR and West Germany also produce selenium but no reliable figures on production levels are available.

Total world production 1,565,423 kilos

Grades available

Selenium is usually made available in the form of powder or granules and most commonly traded as a fine powder (minus 200 mesh) with a purity of either 99.5 per cent or 99.8 per cent minimum.

Production method

Almost all selenium is produced as a by-product of copper production. Selenium-rich slimes are recovered from the anode during the electrolytic refining of copper. To produce the elemental selenium these slimes are treated with sulphuric acid and the metal is then precipitated by a simple chemical operation.

Major uses

The five major uses of selenium are rather sophisticated and it should be noted that, in each, the cost of the metal is small compared with the system or process in which it is used. These uses divide the consumption roughly equally but change slightly according to technical innovation.

It is used in the manufacture of certain types of glass as a decolourizer. In the electronics industry it is used in the manufacture of transformers, semi-conductors and photo-electric cells. Selenium has a unique use in photocopying machines, transferring a photographic image by means of static electricity. The metal is used in combination with cadmium to make an orange/red pigment used extensively in plastics and ceramics. The remaining major use is as an addition to steels, mainly stainless steel, to improve forging characteristics. There are several very minor uses for selenium in the chemical, pharmaceutical, rubber and explosives industries.

Main market features

Selenium has one of the world's most volatile markets. It has all the classic ingredients of volatility: it has a small production, which makes it suitable for market manipulation; the demand for selenium is only fractionally reduced by high prices as explained above; and it is a by-product which sets an upper limit on production.

In times of shortage its value can change by hundreds of per cent only to fall by similar values when consumption declines. Fortunately its production is rather widespread compared with many materials, and no single producer is in a position completely to control prices.

Demand has fallen during this last period of economic recession and prices have fallen to an all-time low. Selenium must be extracted during the refining process of copper ore in which it is found and copper producers seem to be prepared to sell it at the best price they can get, however low. Prices can be expected to rise substantially however *if economic conditions improve.*

Known world reserves

Approximately 80,000 tons.

Method of marketing and pricing

The major selenium producers have recently abandoned their official 'producer price'. They were faced with the dilemma of either having to build up unsaleable stocks (because they could not sell the metal at the high producers' price against competition from Japanese and other producers who were selling on the free market) or joining the free market. Stock levels in both producers' and consumers' hands are very high and it will be some time before they can be reduced sufficiently to reintroduce a producers' price once more.

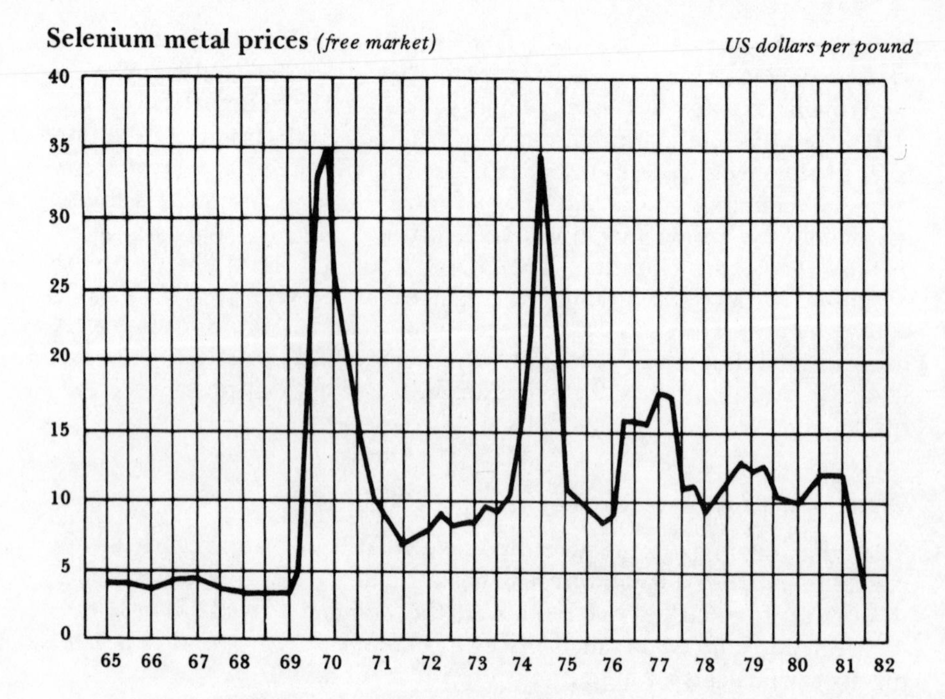
Selenium metal prices (free market)
US dollars per pound
40
35
30
25
20
15
10
5
0
65 66 67 68 69 70 71 72 73 74 75 76 77 78 79 80 81 82

Silicon

Silicon metal production, 1978

ten thousand metric tons

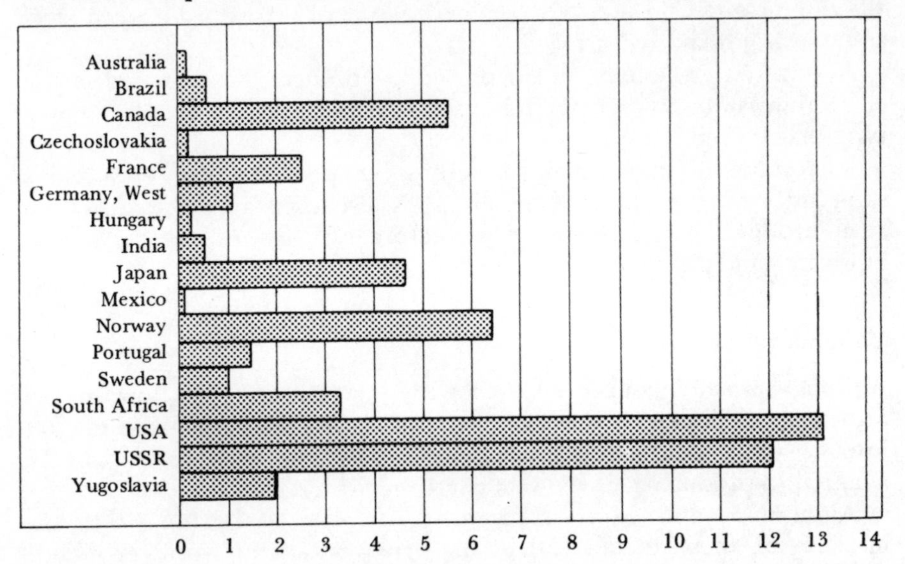

Total world production 549,000 metric tons

Grades available

Most silicon is marketed in the form of irregular lumps usually between 10-100mm in size. The most commonly accepted purity is 98.5 per cent minimum with maximum iron and calcium contents of 0.5 per cent and 0.3 per cent respectively. When silicon is used in the chemical industry, higher purities are often required.

Most silicon is consumed in the form of ferro-silicon, an alloy of silicon and iron. The most popular grade contains 75 per cent silicon.

Production method

High grade quartzite sands and pebbles are the raw material for both silicon metal and ferro-silicon. These are mixed with coke in the production of silicon and with both coke and scrap iron in the production of ferro-silicon, and then smelted in a powerful electrical arc furnace. The raw materials for both products are extremely cheap. The cost of production is mainly the cost of electrical energy. Production plants are usually sited in areas with low cost electricity and often linked to large hydro-electric plants.

Major uses

Approximately 90 per cent of all silicon is produced in the form of ferro-silicon which is consumed in the iron and steel industry as an alloying agent. Consumption is divided roughly equally between steel and cast iron manufacturing.

About 75 per cent of all silicon metal produced is consumed in the aluminium industry where it is used to make aluminium alloys of a type that is used in castings.

Silicon metal (silicon is not strictly a metal in chemical terms and is sometimes referred to as a metaloid) is also used as a de-oxidizer in steel production and in the manufacture of silicones used as waxes, lubricants and polishes.

Main market features

Although we are concerned here with the marketing of pure metals rather than alloys, we must in this case take the ferro-silicon market into account because many plants producing ferro-silicon can be adapted to producing silicon metal and vice versa.

Aluminium and iron ore have such wide applications that the markets of these metals reflect the general state of world economic conditions. As nearly all silicon is used in these two industries it follows that the silicon market is governed in the same way. A higher growth rate for aluminium over steel is predicted which will encourage silicon rather than ferro-silicon production, but silicon production requires more energy to produce than ferro-silicon plus a higher grade of raw material.

Many countries with spare electric power have been attracted to silicon production because the raw material is cheap and relatively common. Electricity is often a difficult commodity to export and production plants are fairly unsophisticated. These countries assume that silicon is like a solid storable exportable form of electrical power.

Prices, however, have not increased even remotely in line with energy prices, mainly because of the severe competition between producers, none of whom is making enough profit to contemplate future increased production. Silicon is an extremely useful metal and demand continues to grow in the aluminium and silicone sector. These factors, combined with the continuing escalation of fuel prices, must mean higher prices, especially for metallic silicon in the future, and may mean a really rapid increase in price in a year or two when demand will almost certainly outstrip the rather low expansion rate of supply.

Known world reserves

Silicon is one of the world's most common elements. Reserves are so enormous that there is no point in estimating them.

Method of marketing and pricing

Ferro-silicon, ferro-manganese and ferro-chrome are known as bulk ferro-alloys and they are traded quite differently from silicon metal. Ferro-silicon is marketed to steel mills in very large tonnages often more than 1,000 tons at a time, either directly by the producers or very often through their agents. There are comparatively few steel mills, so trade is highly competitive, except in times of shortage. There are, however, very many cast iron foundries which buy in quite small quantities from producers' agents and merchants who specialize in the trading and distribution of bulk ferro-alloys. Most of these merchants are small, nationally based companies, but there are

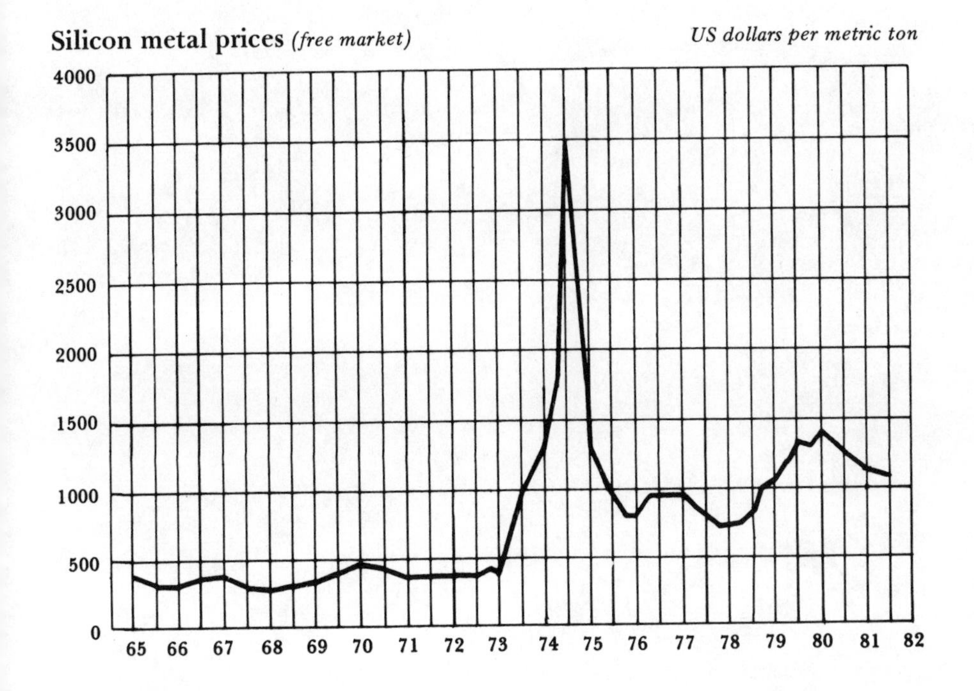

several very large international merchants trading in larger tonnages who often have agency agreements with the producers.

There are comparatively loose forms of producer price control but most trade is subject to supply and demand.

Silicon metal is also distributed directly by producers or their agents at a ruling producer price, but there is a large and active free market in the metal. The Scandinavian producers are the most successful at maintaining a producer price. Merchants in silicon metal are usually different from those trading in ferro-silicon. The material they deal in comes mainly from Yugoslavia and Spain, and from that which is offered for export by producers outside the country of production.

Sodium

Sodium metal production, 1978 *ten thousand metric tons*

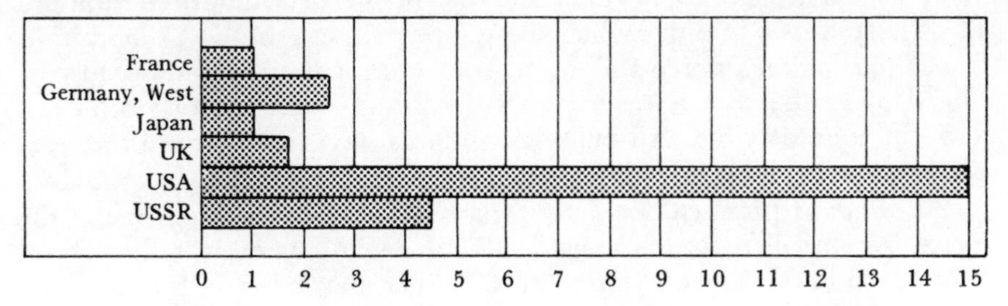

Total world production 247,000 metric tons

Grades available

Sodium is an extremely soft metal which reacts violently with water. It is usually sold in the form of sticks or blocks which have to be protected from water in the atmosphere by submerging in oil or by placing in airtight drums.

Production method

Sodium is produced electrolytically from molten sodium chloride (common salt).

Major uses

About 80 per cent of production is used to make the anti-knocking agents tetraethyl and tetramethyl lead used in petrol. Its other major uses are in the manufacture of titanium metal (magnesium metal can also be used for this purpose) and as a coolant in nuclear reactors.

Main market features

There is very little merchant activity in sodium. This is mainly because there are few producers in the world, and the main producers also manufacture the anti-knocking compounds, which are the biggest outlet for this metal. The USSR does export material from time to time but only when in surplus to its own requirements.

Known world reserves

Reserves of raw material are only as limited as the salt in the ocean.

Method of marketing and pricing

The market price of sodium is based only on its cost price. The raw material costs next to nothing and the main costs of production are electricity and handling, which have been rising rapidly. Competition is not particularly fierce for the reasons outlined above and because of the danger of moving the material over long distances, or storing for long periods of time. All major consumers buy direct from producers who sell at a similar price in the same market.

As there is practically no free market activity in sodium metal the price is controlled by the major producers who currently sell at about 50¢ per lb in the United States and £750 per metric ton in Europe.

These prices have risen by approximately 10 to 15 per cent per annum over the last three years.

Tantalum

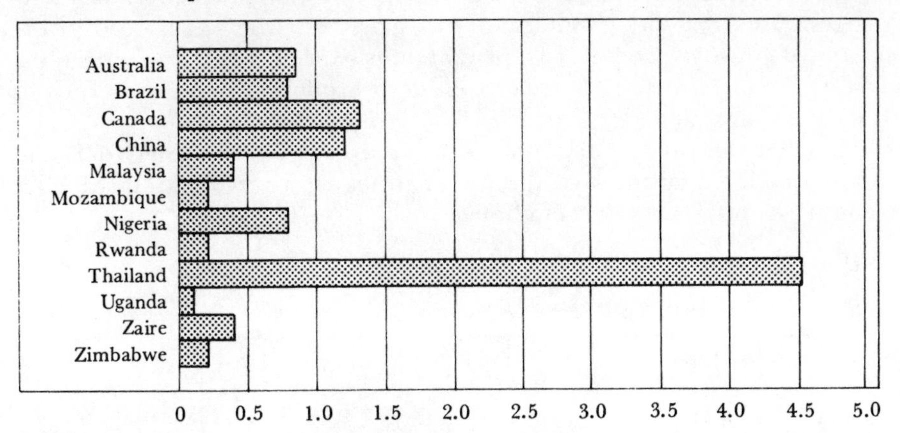

Total world production 1,230 metric tons

Grades available

Most tantalum is traded in the form of its ore, tantalite, which is sold in two main grades: that containing a minimum of 60 per cent Ta_2O_5; and that containing approximately 30 per cent Ta_2O_5.

There is very little trade in metallic tantalum, but the producers' main products are powder, sheet and bar of 99.9 per cent minimum purity.

Production methods

Tantalum oxide is extracted from its natural occurring ores, tantalite and columbite (the two often occur together). The oxide is first concentrated by magnetic separation and the metal is then produced by electrolytic means.

Major uses

Like tungsten and columbium, tantalum is one of the refractory metals. It has a high melting point and is very resistant to corrosion. In the past it was added to steel to make alloys similar to those made with columbium. The discovery of adequate deposits of columbium ore which is less dense than tantalum has almost eliminated the use of tantalum as a steel additive.

Tantalum's corrosion resistance has found many applications in the chemical industry where it is used to make pipes, crucibles, retorts, etc.

The electronics industry is tantalum's biggest consumer. Here its largest use is in the manufacture of capacitors, where it is used in the form of powder produced from tantalum oxide by first converting the oxide to the fluoride. It is also used to produce components such as contact points and electrodes.

Small quantities of tantalum carbide are used in cutting tools and even smaller quantities of the metal are used in surgery to make clips and struts to repair veins and bones.

Tantalum ore prices *(free market)* *US dollars per lb Ta_2O_5 contained*

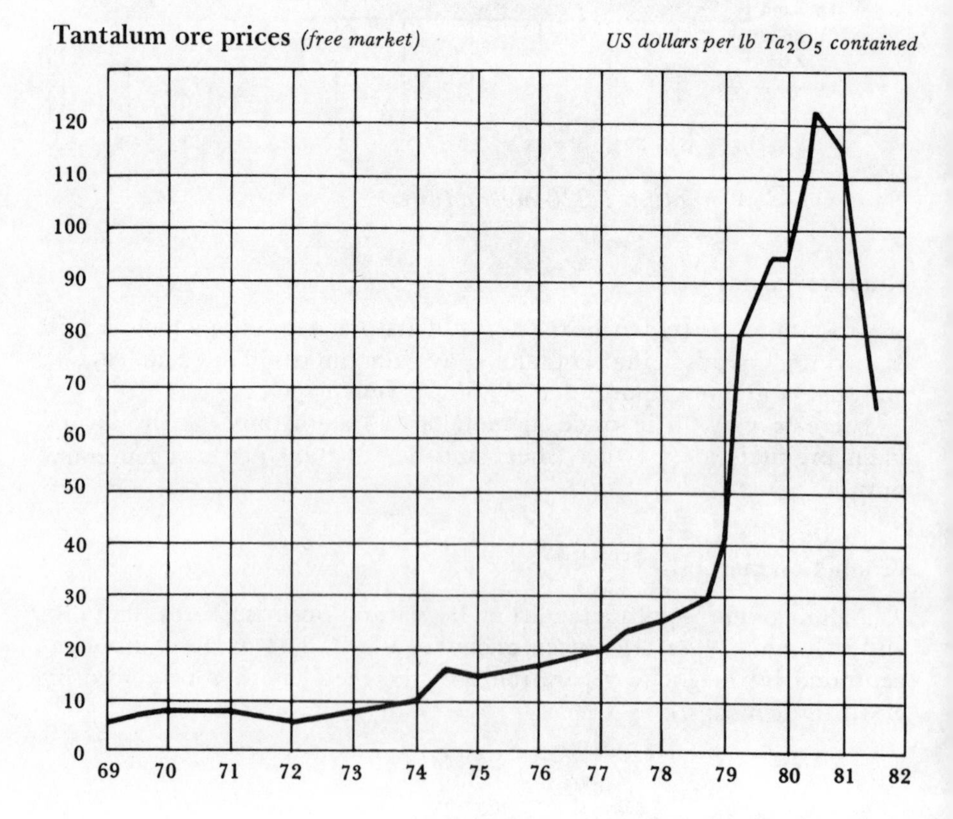

Main market features

Demand for tantalum for its use in capacitors increased considerably and prices reached such a high level in 1980 that the metal became uncompetitive in most of its other uses. High prices have not only encouraged competition from other metals but have also stimulated increased production. These factors have helped to bring down the price. Demand for tantalum capacitors remains high, however, and one cannot expect prices to fall back very much more in the foreseeable future.

Method of marketing and pricing

Tantalum ore has an important application for making refractories (used for furnace linings, etc) outside its use as a raw material for making tantalum.

All tantalum metal producers maintain a close relationship with consumers and are often required to make the semi-fabricated parts that consumers require. These products may have a strict degree of technical tolerance which is reflected in the price.

Most merchant activity is restricted to the trading of tantalum ore but some merchants trade in scrap tantalum, which is usually recycled back to the producers.

Tellurium

Tellurium metal production, 1979* *ten thousand kilos*

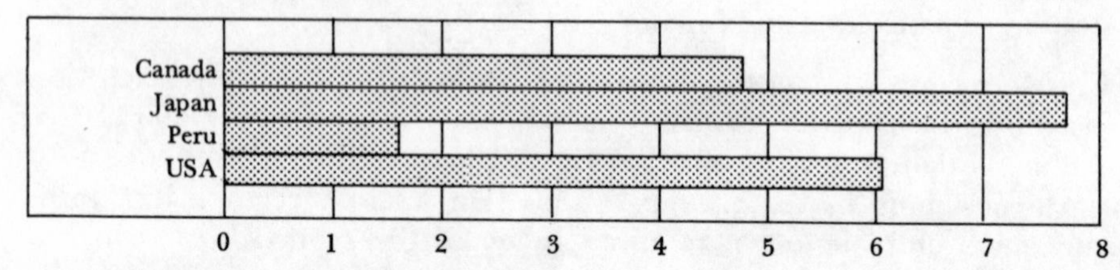

* Australia, Belgium, the USSR and West Germany also produce tellurium but no reliable figures on production levels are available.

Total world production approximately 220,000 kilos

Grades available

Tellurium is usually marketed in the form of small ingots or slabs, or in the form of fine powder. The purity of commercial grade material is normally between 99.7 per cent and 99.9 per cent minimum.

Production method

Tellurium is only produced as a by-product of other metals, particularly copper, but it may be present in commercially recoverable quantities in certain zinc and lead ores. Tellurium containing slimes are recovered from the anode during the electrolytic refining of copper. These slimes are leached with caustic soda. Tellurium oxide is then precipitated from the solution by acidification and the oxide is roasted to the metal and refined.

Major uses

Tellurium is used as an alloy addition, in very small proportions to certain steels and copper alloys. In both cases the tellurium content improves the machinability of the alloy. It may be added to iron and steel in the form of ferro-tellurium containing about 50 per cent tellurium, the balance is iron.

Its addition to steels reduces porosity and acts as a grain refiner.

The compounds have various uses in the chemical industry in ceramics, rubber vulcanization and in detonators and pesticides. However, these chemical uses only amount to about 10 per cent of the tellurium market.

An alloy of bismuth and tellurium displays unique electro-thermic characteristics, but no major industrial application has yet been found for this property.

Main market features

Tellurium's uses are of no particular importance and all have known substitutes, but tellurium is preferred to other products at current commecial differentials.

Only about half of the potential supply of tellurium in copper ore is utilized at present, due to the cost of its extraction. Any future increase in demand could be met quite easily by improving extraction techniques.

Prices are likely to keep in line with the increased cost of production, but any future growth in demand may lead to a temporary price rise while production is being increased.

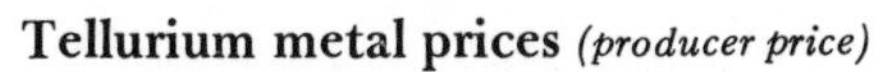

Tellurium metal prices *(producer price)* *US dollars per pound*

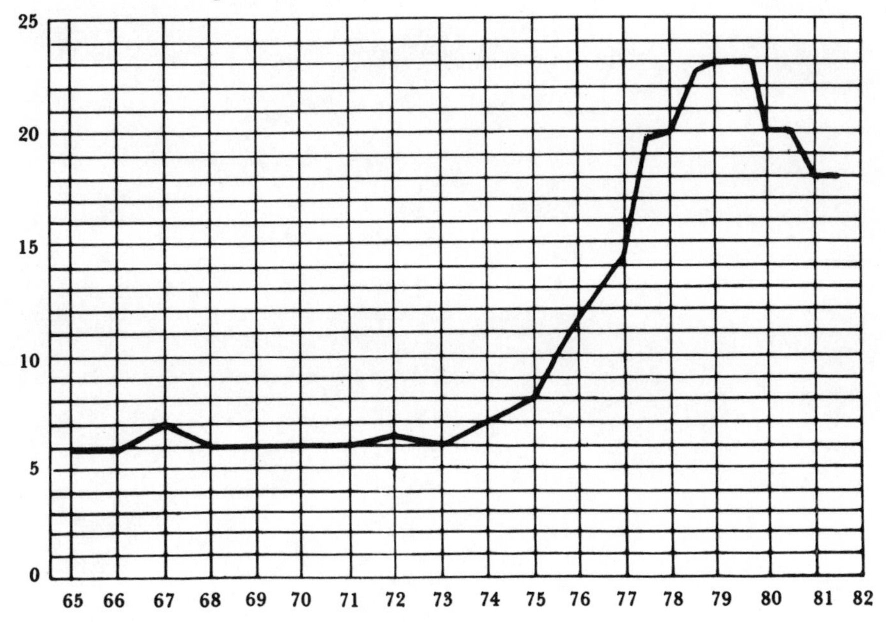

Known world reserves

Approximately 50,000 tons of tellurium metal.

Method of marketing and pricing

The tellurium market is very small and its supply/demand characteristics have not attracted much merchant activity. The major producers handle most of the marketing directly or through agents. Prices are fixed by producers, but some business is done on a free market basis, especially that between producers and state-owned buying corporations, often with a merchant as intermediary.

Tin

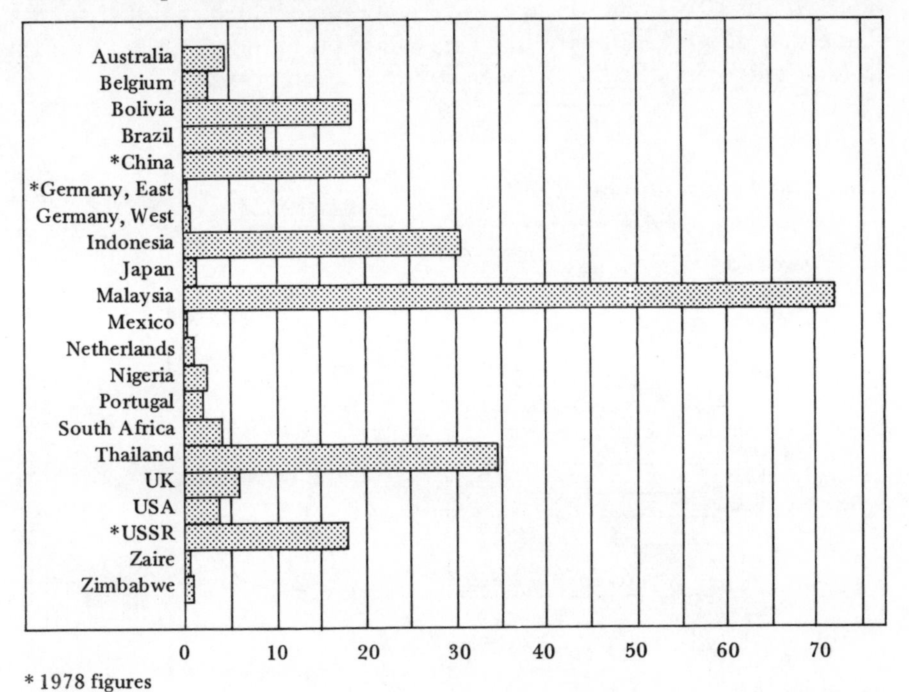

Total world production 228,000 metric tons

Grades available

Tin is sold on the international market in ingot form (up to 50kg each) with a purity of 99.75 per cent or at a premium price for high grade material with a purity of 99.85 per cent minimum.

Production method

Tin's main ore, cassiterite (tin oxide) is mainly found in alluvial deposits in river beds and deltas and recovered by dredging methods. It is also recovered by the gravel pump mining method, and by traditional mining in such areas as Bolivia.

The ore, either from dry or wet deposits, is usually concentrated by magnetic or electrostatic means and the concentrate is smelted with coke in a reverberatory furnace to reduce it to metal.

Major uses

The major user of tin is still the canning industry. Tin cans, used for packaging food and drink, are made of steel sheet, plated with a thin layer of tin. This material is known as tinplate which is rolled from a large slab of steel clad with tin. Tin is an important constituent of most solders and other alloys such as babbit metal and bronze.

Tin refined consumption, 1980 *thousand metric tons*

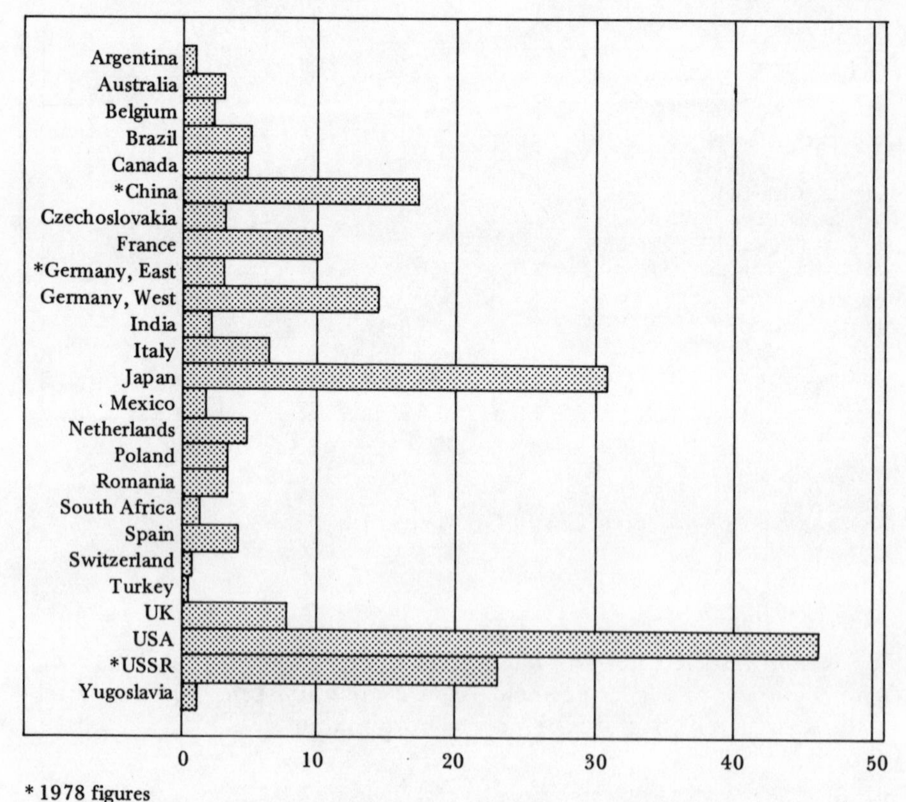

* 1978 figures

Total world consumption 222,000 metric tons

Main market features

Most of the world's exported tin is produced in the geographically proximate countries of Indonesia, Malaysia and Thailand, and in Bolivia. Bolivian tin has to be mined from hard rock in inaccessible areas of the Andes mountains whereas the material from South-East Asia is produced by simple mining and dredging techniques. Tin from the latter source is therefore cheaper to produce and the income of Bolivian miners has to be kept low in order to compete. Very little tin is produced in the major consuming countries.

These factors have allowed the tin market to be controlled with more success than has been experienced by the producers of most other important industrial metals. This has been achieved in the past by the agency of the International Tin Agreement (ITA) to which most major producing and consuming countries (with the notable exception of China) have been signatories.

In 1981 the United States, the largest single consumer of the metal, failed to agree to the terms of the new ITA and will instead attempt to control prices by using the vast stocks of tin which it has accumulated in its strategic stockpile. These stocks alone could supply US needs for several years but will not be enough to maintain low prices indefinitely. In the meantime producers seem to be determined to raise tin prices and, unlike most metals, the price of tin has increased during the present recession.

Tin is used mainly in the canning industry and an upper limit can be computed for the price of tin while substitutes for tin-plate such as aluminium, plastics and glass can be used to preserve and pack food and drink.

Method of marketing and pricing

Tin is traded on the London Metal Exchange and on the Penang tin market and almost all of the world's tin trade is based on the price of one or other of these markets or a mixture of both. In the Penang market buyers present prior bids where each buyer stipulates the amount of tin he wants and the highest price he is willing to pay. The market officials then pick the highest bids that total the quantity of ore concentrates delivered to the market by the miners that day and the price of the lowest of these bids is used to price all the sales contracts. The International Tin Council employs an official called the buffer stock manager who attempts to control the tin market according to the terms of the International Tin Agreement. He is obliged to buy tin from the market if the price drops below an agreed limit and must sell tin from his 'buffer stock' if the price rises above another agreed limit. The funds he requires to buy tin are supposed to be provided by both producers and consumers but consumers are often reluctant to increase the price of a commodity they buy. In the past the United States has offered tin from its stockpile rather than cash, which has not always been acceptable to producers. Now that the USA seems unlikely to continue as a signatory of the ITA the producers may be in a stronger position to control the market by themselves.

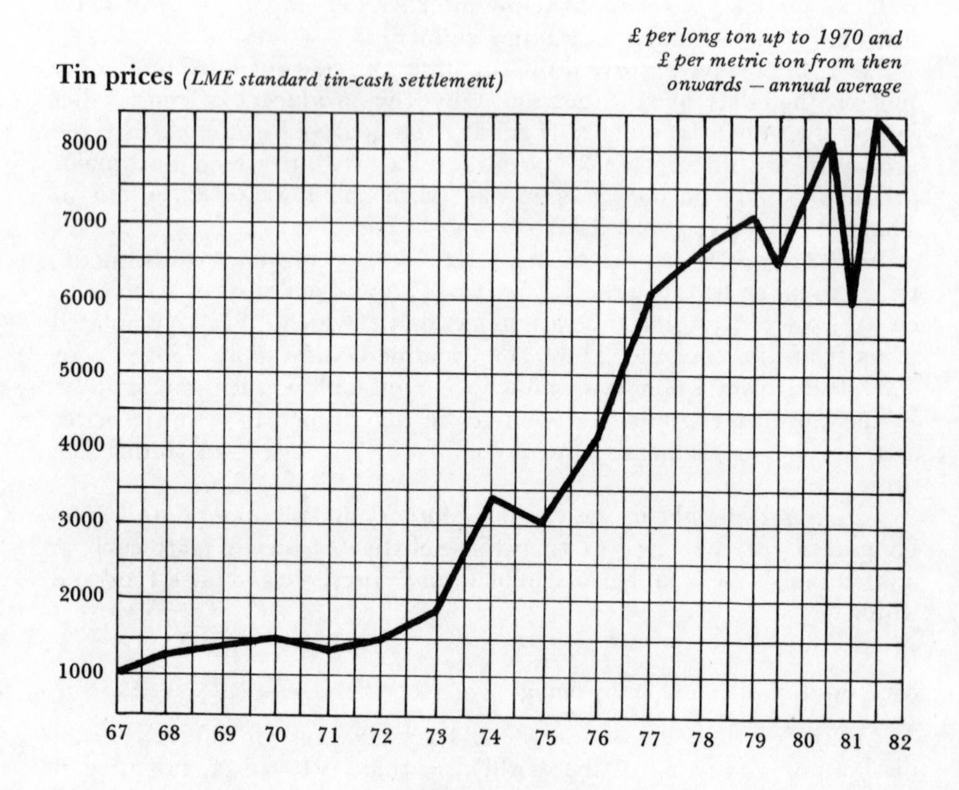

Titanium

Titanium sponge estimated production, 1980 *thousand metric tons*

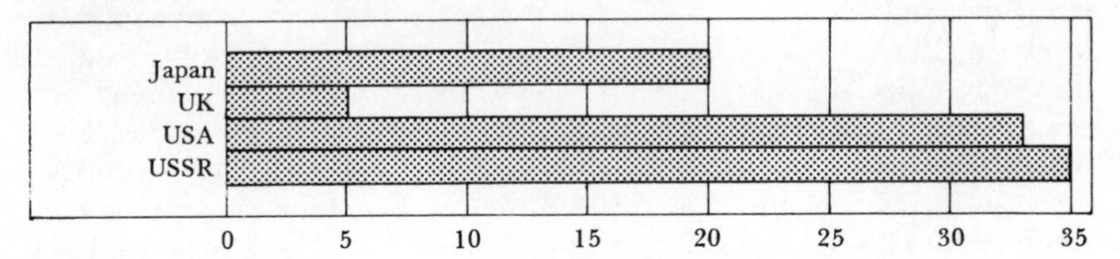

Total world production 92,900 metric tons

Grades available

Titanium's uses are of a highly technical nature. Many producers are therefore obliged to make titanium in several sizes and forms, that is slabs, sheet, bars, wire and other semi-fabricated types, but pure titanium is traded mainly in the form of sponge (irregular granules) or briquettes (compressed blocks of sponge). This material is usually of 99.6 per cent minimum purity.

Production method

The metal is produced by first converting the ore, rutile or ilmenite into titanium dioxide by means of a simple chemical treatment. The dioxide is then converted to the tetrachloride and heated with magnesium or sodium which reduces it to the metal.

Major uses

There are many independent industrial uses for both the ore and the oxide. Titanium metal's physical properties, its lightness, strength and resistance to corrosion have resulted in many new uses for the metal in modern times. It has been found particularly useful in the aerospace industry, where it is used for such items as turbine blades, engine shafts and in the framework and skins of spacecraft and high performance aircraft. Its resistance to corrosion makes the metal suitable for applications such as pipes, pumps and containers in the chemical industry and in electro-plating. The metal has found an increasing use as an alloy addition in certain steels, some of which are used in automobile silencers (mufflers). It is added to steel either as pure titanium or in the form of ferro-titanium, a master alloy which is made by melting scrap titanium and iron, or by direct reduction of the ore with iron.

Main market features

The boom in both civil and military aircraft production created enormous demand for titanium which caused prices to reach unprecedented levels in 1980. New production capacity has been quickly installed, however, especially in Japan and the USA to meet this demand and prices have since dropped considerably. Supplies of raw material for making titanium are plentiful. Although one cannot expect a further increase in price in the near future, free market trade is likely to continue to be very active as larger tonnages are now being produced and consumed. Titanium production is confined to a very few manufacturers in industrialized countries. These countries are also the major consumers of this metal.

Titanium (sponge) metal prices *(free market)* *US dollars per kilo*

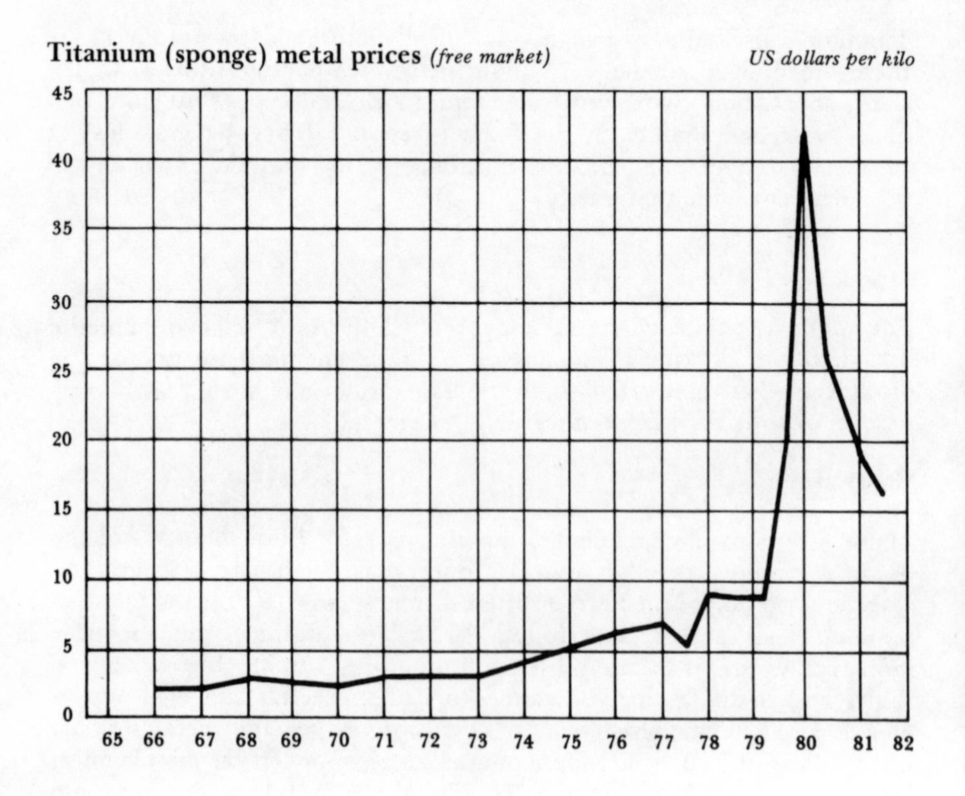

Known world reserves

Titanium metal in ore is approximately 4,350,000 metric tons. However, titanium can be extracted from other minerals less competitively.

Method of marketing and pricing

The producers of titanium control its price in the country of production. Their price is limited only by the price of imported titanium products from other countries. There is, however, an active free market trade in metal originating from the USSR, usually in the form of sponge or briquettes. Some quantities are also available from Japan and very recently from China in the form of sponge. Scrap titanium is a more interesting market, however. This material arises in the form of machine turnings, sheet cuttings or solids. These come mainly from large aircraft manufacturing centres, such as the USA, the USSR and West Germany. Prices vary according to the alloy type, form and contamination.

Tungsten

Tungsten ore production, 1978 *thousand metric tons WO_3 content*

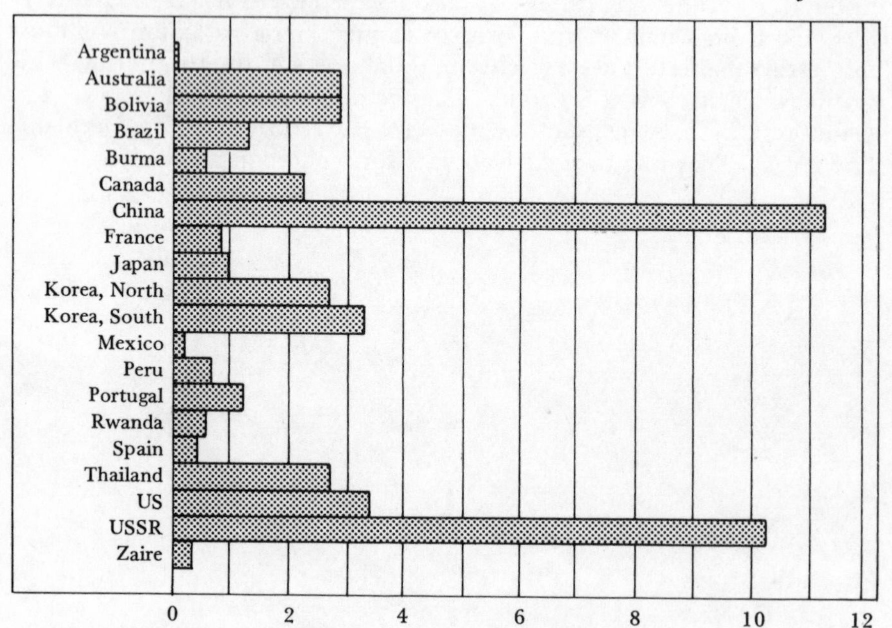

Total world production 42,774 metric tons

Grades available

Tungsten is traded mainly in the intermediate forms of the metal. These are the concentrates, wolframite and scheelite, ferro-tungsten (the alloy of iron and tungsten) and ammonium paratungstate (APT). Pure tungsten, usually in the form of powder, is also traded, but this is a very minor proportion of the business.

Up to 90 per cent of the world's international business in tungsten is conducted in the concentrates. These are traded in metric ton units. One MTU is a single per cent of WO_3 contained in a metric ton of the concentrate, i.e. 22.046 lbs of WO_3. Most commercial grades of tungsten concentrates contain about 65-70 per cent WO_3 which means that they have 65-70 MTU per metric ton.

The concentrates traded are scheelite ($CaWO_4$), ferberite ($FeWO_4$), huebnerite ($MnWO_4$) and wolframite ($[FeMn]WO_4$). Ferberite, huebnerite and wolframite are loosely referred to as wolfram ore or black ore in the trade. Traders pay particular attention to the levels of impurities in these concentrates. The most important of these are tin, arsenic, phosphorus, sulphur, copper and molybdenum. The average scheelite or wolfram ore would contain:

WO_3 (tungsten oxide)	68 per cent
Sn (tin)	0.2 per cent maximum
As (arsenic)	0.2 per cent maximum
P (phosphorus)	0.04 per cent maximum
S (sulphur)	0.5 per cent maximum
Cu (copper)	0.1 per cent maximum
Mo (molybdenum)	0.3 per cent maximum

To confuse the issue further a product called synthetic scheelite is regularly traded which is chemically produced from low grade ores. It is less popular than natural scheelite due mainly to the higher import tariffs it carries into some consumer countries.

Production method

All tungsten concentrates are produced by very simple flotation and gravitational separation from the ore. Ferro-tungsten is either produced by the normal alumino thermic method (reduced from the ore with aluminium powder in the presence of iron) or by reduction in an electric arc furnace. Tungsten scrap is often the feedstock in this last method.

Most pure tungsten is produced in powder form from APT by reduction with hydrogen. APT is produced chemically from the concentrates.

Major uses

There are very many tungsten-containing alloys. Some are steels (scheelite can be used as a direct addition to steel for making many of these alloys) with wide application in all forms of machine tools and mining equipment. Tungsten carbide alloys, made by sintering tungsten powder with other metal powders, also have very wide applications for tools, machinery, components, bullets and tyre studs, etc.

Fine tungsten wire is used for electric lamp filaments, and tungsten metal or alloys are used as contact points in electrical switch gear.

Tungsten compounds are widely used in the chemical industry in lubricants, pigments, catalysts and in many other minor applications.

Main market features

Most of the uses for tungsten are vitally important for the world's industrial activity. Substitution on a large scale with other materials in these uses is very difficult. Fluctuations in demand usually reflect the general state of the world's economic situation, but the price trend in the last two decades has been rising substantially due mainly to the increase in demand and to growing control by tungsten producers.

Tungsten is a typical example of a vitally important raw material which is produced mainly in third world countries but consumed mainly in industrialized countries. It is used widely in the production of armaments and armour plating, and the forecast international increase in expenditure on armaments should ensure a significant increase in demand for tungsten.

China alone produces over a quarter of the world's supply and is in a position to exercise a massive degree of control over prices. The world's producers display a degree of cooperation but the smaller producers in South East Asia and South America rely upon China to set the pace.

With almost no substitution possible, it is reasonable to suppose producers will exert control over supplies to ensure that prices continue to rise, especially as China is expected to increase its internal demand for tungsten over the next decade.

Known world reserves

Reserves of tungsten ore (wolfram content) are approximately 1,200,000 metric tons. 900,000 metric tons are estimated to be in China.

Method of marketing and pricing

The tungsten market, perhaps more than any other, is conducted by international merchants.

It is extremely volatile and very speculative. The purchases from China and the sales to Eastern European countries, including the USSR, are the backbone of the trade. More trade in tungsten is being conducted between China and the USSR these days, however, in spite of some political antagonism between the two countries.

The Chinese conduct a responsible, if somewhat erratic, sales policy. They may withdraw entirely as sellers for long periods of time if they feel the price is too low, only to appear as aggressive sellers when the price suits them or when they have a requirement for foreign currency. Their No 1 grade wolfram ore specification is a standard for world trade.

The other smaller producers have their own specification for their product, and assays usually have to be made by independent assayers on the outturn of cargoes.

With the exception of North America, very few producers sell directly to consumers.

The most authoritative price for the ore is recorded in the London Metal Bulletin, but there is much bitter controversy from producers, consumers and merchants alike concerning the accuracy of these records.

Many efforts have been made to find a more suitable way of international pricing but without success.

In 1978 an International Price Index, using information drawn from both consumers and western world producers, was introduced as a means of reflecting more price influences than just the free market merchant price published by the Metal Bulletin.

Tungsten is an expensive product with a volatile market which has attracted intense speculation, often with dire results for the speculator. Without this speculation, however, it would be difficult to envisage how

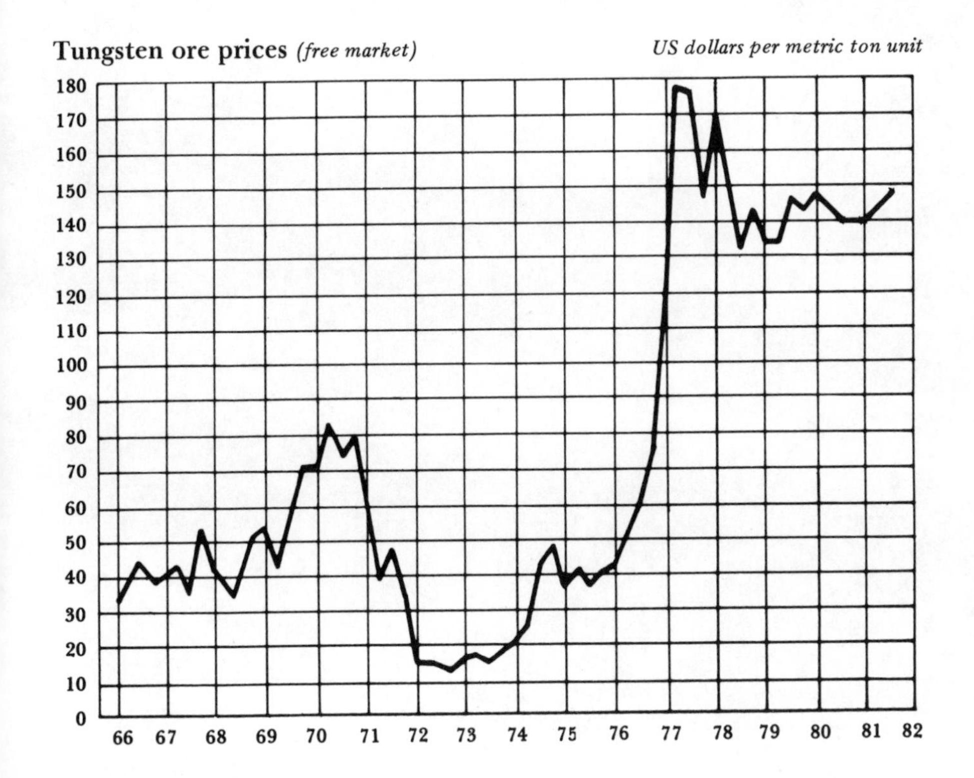

the tungsten market could be conducted.

The General Services Administration (GSA), the American strategic stockpile, retains large stocks of tungsten in various forms. This material is currently made available to buyers in regular official sales. It is an additional confusing ingredient in this very complicated market.

Vanadium

Vanadium ore production, 1979 *thousand metric tons*

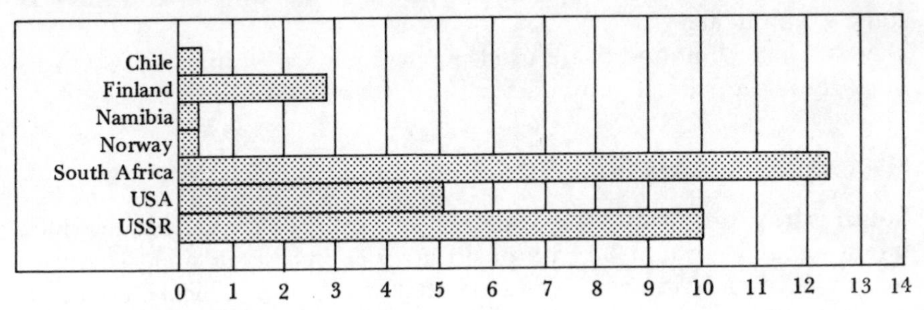

Total world production 37,568 metric tons

Grades available

The pure metal is seldom traded, as vanadium additions to steel are usually made in the form of ferro-vanadium. The bulk of trades are in vanadium pentoxide (V_2O_5) or ferro-vanadium usually containing either 80 per cent or 50 per cent vanadium, the balance is iron.

Production method

Although vanadium is a very common metal in the earth's crust, it is only rarely found in commercially viable concentrations, often in association with other ores and minerals.

It is also extracted from the flue dust of power stations which use vanadium rich cokes and coals.

Ores or vanadium-containing slags are converted into vanadium oxide or directly into ferro-vanadium.

The raw material is concentrated, leached and roasted to the oxide.

Ferro-vanadium is usually produced aluminothermically by reduction of the oxide in the presence of aluminium powder and iron.

Major uses

Most vanadium is used in the form of ferro-vanadium in the production of structural steels. An important growth area for such steels is in the manufacture of oil and gas pipelines. The growth in production of pipelines has become particularly marked since the tremendous rise in fuel prices in 1974 and 1975.

There are important uses for such steels in bridge building and high-rise buildings.

Other types of steels containing vanadium are used wherever toughness is required in machinery components, mining equipment and tools.

Certain non-ferrous alloys incorporate a vanadium content, notably some titanium alloys.

Vanadium compounds are used in the chemical industry as catalysts in glazes and for the manufacture of some synthetic rubbers.

Main market features

South Africa produces approximately half the world's vanadium which implies that it should be in a position to exercise a degree of control on the market. Although it is true that the South Africans could set the pace for price changes, they appear to be conscious of the fact that the other so-called refractory metals, tungsten, molybdenum and niobium, are able to replace vanadium in many of its applications.

Ferro-Vanadium prices *(free market)*

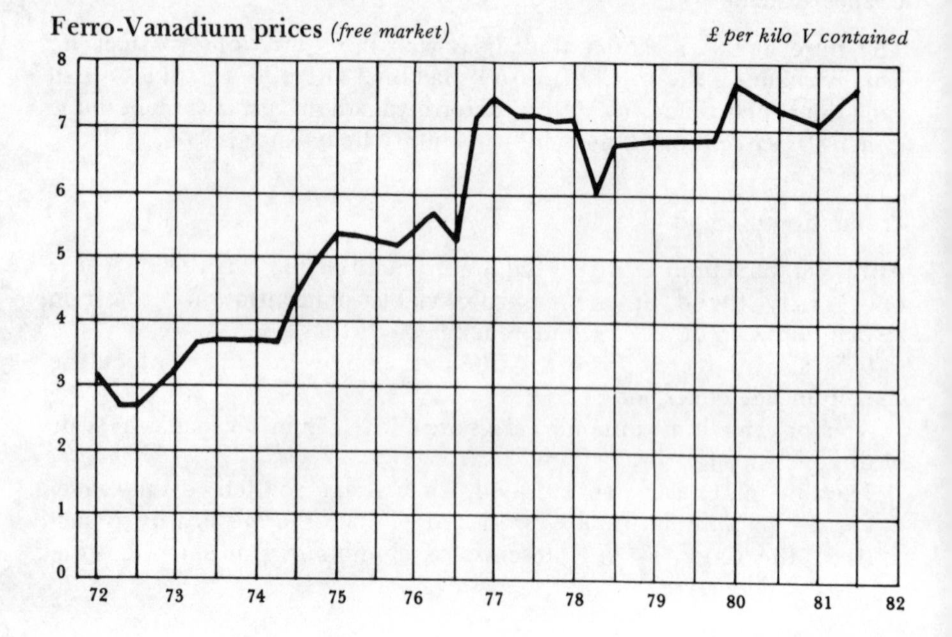

This feature of the market perhaps more than any other has kept price movements reasonably gentle.

The current boom in pipeline construction has created an increased demand for vanadium and its competitors which may signal a change in this historic pattern.

Very recently the Chinese have appeared as a substantial seller of vanadium pentoxide, but as yet it is unclear whether these quantities will continue to be made available to the Western market and what effect this might have on prices.

Any long-term shortage is likely to tempt the few major vanadium producing countries to flex their market muscles.

Method of marketing and pricing

The influence of merchant activity is only felt in times of shortage. The major producers normally have a good deal of success in controlling prices by using an official producer price when contracting with consumers, often on a long-term basis. When producers decide to or are forced to restrict sales due to increased demand, merchants are ready to trade at a premium to the official price. Other sources of material during these times would come from surplus consumers' stocks and other smaller producers.

Zinc

Zinc metal production, 1980 *hundred thousand metric tons*

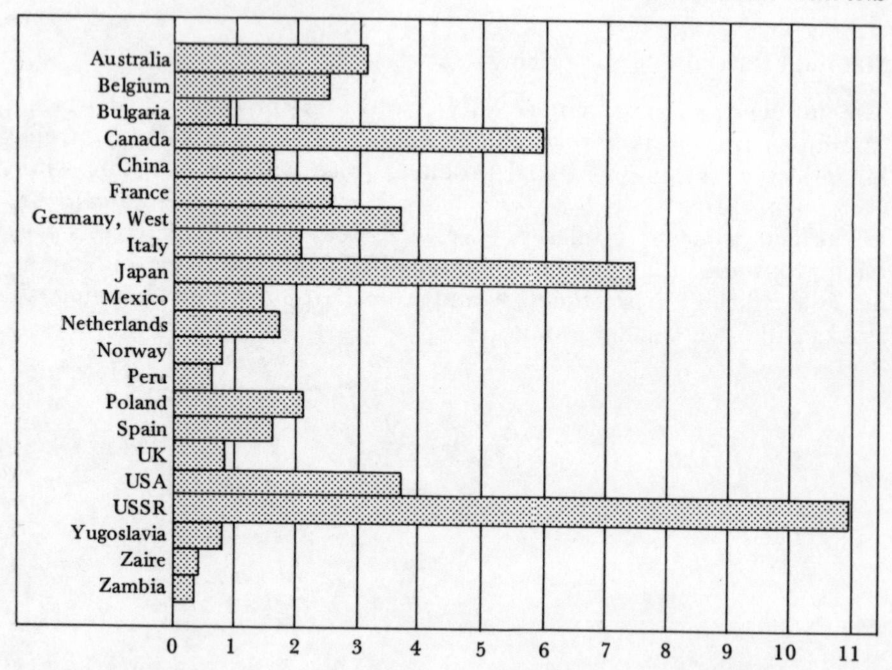

Total world production 5,800,000 metric tons

Grades available

The most commonly traded forms of zinc are 'electrolytic' with the purity of 99.95 per cent minimum, 'GOB', standing for good ordinary brand, which is 98.5 per cent minimum, usually debased with lead, and special high grade with the purity greater than 99.99 per cent. All these types are traded in the form of ingots.

Production method

There are a number of different zinc-containing ores but the most commonly commercialized deposits are sulphide ores and the less common oxide ores.

Zinc ores often occur in combination with lead ores and with minor quantities of rarer metals, such as cadmium and indium. These other metals are separated and extracted during the refining process.

The sulphide ore is first roasted to remove sulphur. This sulphur is used to make sulphuric acid which can then be used to dissolve the zinc contained in the ore, to produce zinc sulphate. The zinc in turn is deposited from the solution onto cathodes in the electrolytic refining process. Zinc can also be refined by distillation, where the ore is roasted with carbon and the distillate collected. In both methods the refined metal is cast into slabs.

Zinc metal consumption, 1980 *hundred thousand metric tons*

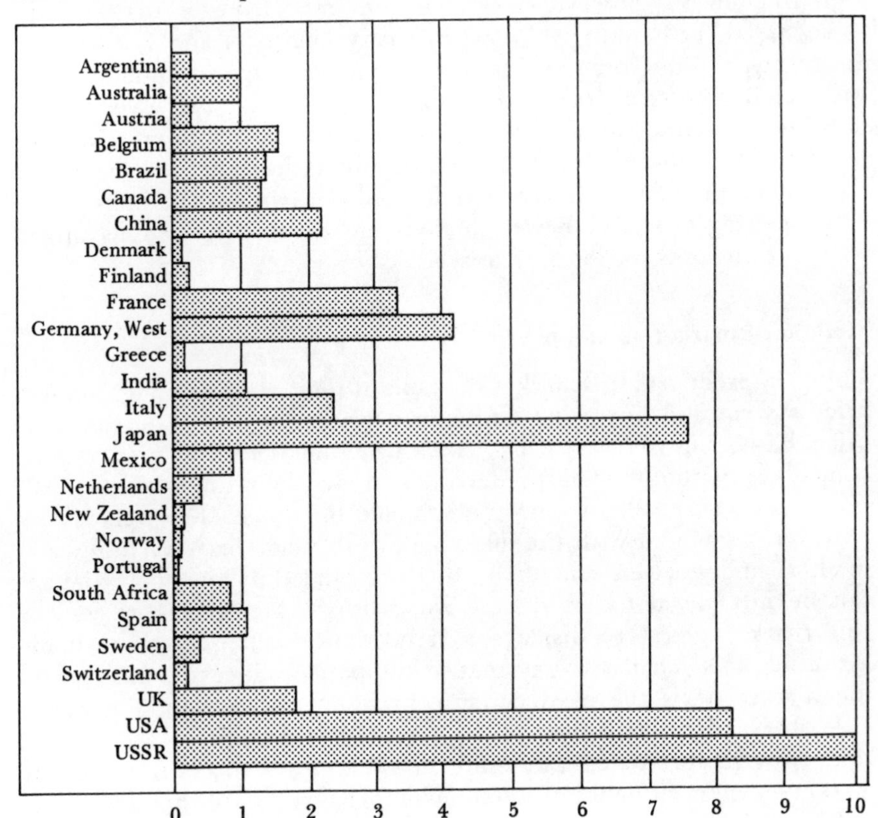

Total world consumption 5,700,000 metric tons

Major uses

One of zinc's most important uses is in galvanizing, the process of covering steel with a layer of zinc for protection from corrosion. This is commonly used to protect roofing sheets, girders, nails, ladders and buckets etc, and can be achieved by dipping the item in the molten metal or by electrolytic plating. Another major use is in brass, its alloy with copper.

Zinc and zinc alloys are also used extensively to produce die-casting alloys where they compete fiercely with aluminium. Zinc is heavier, but aluminium more costly. Zinc sheet has many uses, the most important being the cases for dry-cell batteries. Zinc oxide is used extensively in the rubber industry and in making white paint and pigments.

Main market features

Like lead, much of the world's zinc is produced in the countries in which it is consumed. A great deal of the zinc metal produced in Western countries, however, is refined from imported ore. An extremely toxic metal, cadmium, is found in many zinc ores and is extracted during the refining process. Many zinc smelters are not equipped with sufficient protective devices to ensure that no cadmium escapes into the environment. For this reason some smelting plants especially in the USA have been closed down and others have needed expensive modernization. This has resulted in a slowdown in the increase of zinc smelting capacity. Fewer smelters are capable of refining mixed lead and zinc ores for the same reason.

Method of marketing and pricing

Major US producers still market the bulk of their product on a producer price system and buy what zinc concentrates they need on the same price basis, but many smaller smelting companies and zinc mining companies without smelting facilities trade their material on LME prices. The turnover of zinc trades made on the LME is very small, however, compared with the huge bulk deals made between mines and smelters and smelters and their customers and this 'marginal' tonnage can be influenced by producers who wish to see higher prices. The zinc market therefore displays a trend which can be seen in many metal markets. That is to say that in the past producers could control the market using the producer price system. This system was rather inflexible and broke down but so far only in Europe in the case of zinc. Now the producers can still exercise a degree of control over the market by their influence on the much more flexible free market.

US producers can never allow their producer price to stray too far from the free market price: if they pitched it too high zinc would flood in from outside the USA and if their price was too low they could not buy concentrates competitively, so the US too is being drawn into the free market system.

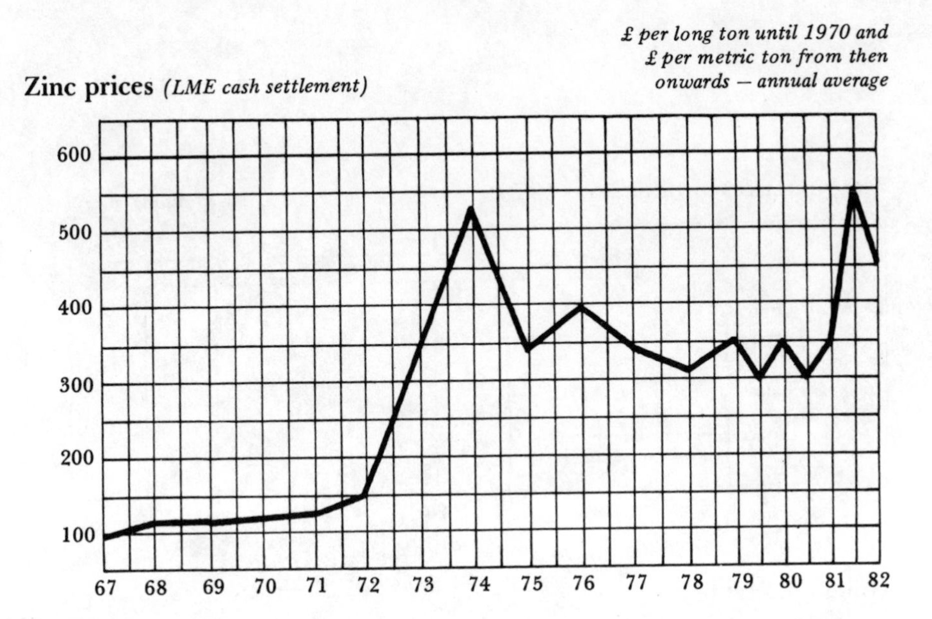

Zirconium

Zirconium ore production, 1979

ten thousand metric tons

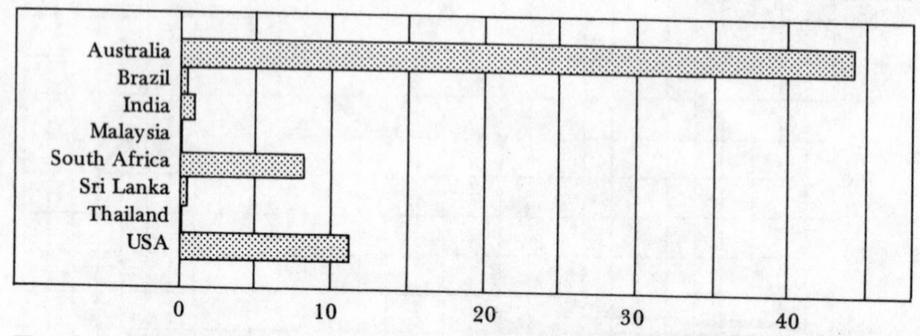

Total world production approx 630,000 metric tons

Grades available

The most common zirconium containing substances are its alloys, some of which are known as zircaloy, containing small quantities of tin or tin with iron, chrome and nickel. Zirconium metal is usually sold in the form of sponge and ingot.

Production method

Zirconium's ore, zircon, is roasted with coke in an electric arc furnace which produces zirconium carbonitride. This is then chlorinated to produce the chloride which is in turn reacted with magnesium metal to produce zirconium metal.

Major uses

Zircon, the ore, has a major use in its own right for making foundry moulding sands.

Several zirconium compounds have uses for making waxes, tanning agents, rust and water repellents, and deodorants.

The alloys have uses in the manufacture of pipes, pumps, valves and fittings in the chemical industry where their anti-corrosion properties are required.

Some pure metal is used in the construction of components for nuclear reactors and for making foil used in flash bulbs.

Main market features

Zirconium is produced by very few manufacturers. Its raw material is plentiful and very little of it is used to produce the metal. It is rather expensive to produce when one considers that another metal, magnesium, is required to make it. Its price has been rising steadily over the years in line with production costs. The price of its rather inexpensive ore, however, has been changing somewhat more erratically due to the changing fortunes in the markets of its non-metallic uses. This all makes the market rather unexciting for metal merchants who confine their activity to trading in the ore or the scrap alloy.

No drastic upsurge in demand is foreseen but a major expansion in nuclear plant construction could cause a temporary lack of production capacity.

Method of marketing and pricing

The few manufacturers that there are sell directly to consumers. They also offer a range of alloys and in most cases semi-fabricated products as well. These producers determine what price they can charge and competition between them is not particularly fierce.

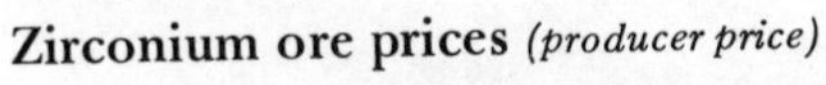

Zirconium ore prices *(producer price)* *£ per metric ton*

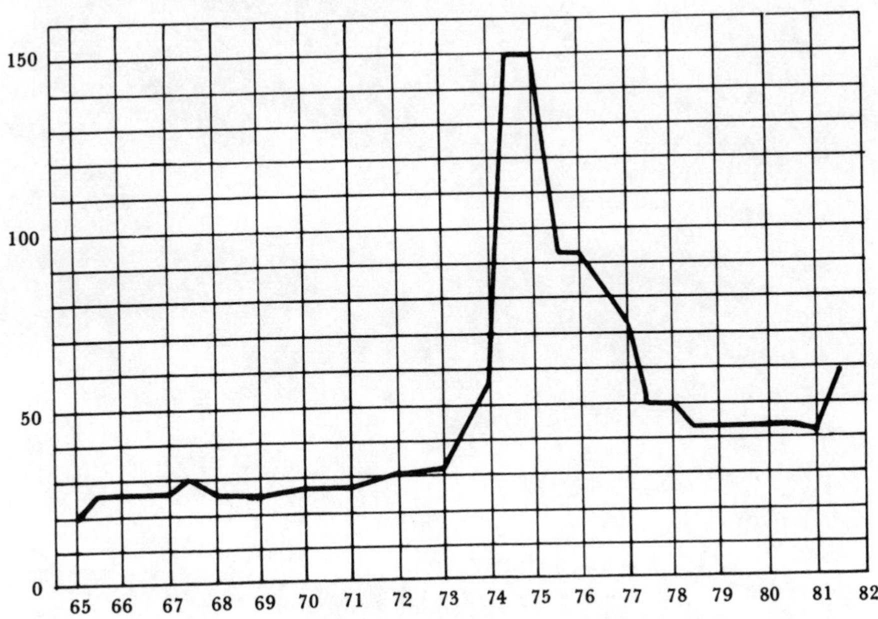

Appendices

Appendix 1:
Useful Addresses

Associations and information centres

The Aluminium Federation
Broadway House
Calthorpe Road
Birmingham B15 1TN
United Kingdom
Tel 021 455 0311

American Bureau of Metal Statistics Inc
420 Lexington Avenue
New York
New York 10017
USA
Tel 212 8679450
Telex 147130
Cables Amburostat
President William J Lambert

American Society for Metals
Metals Park
Ohio 44073
USA
Tel 216 3385151
Telex 980619
Managing Director Allan Ray Putman

The Australasian Institute of Metals
Federal Headquarters
191 Royal Parade
Parkville
Victoria 3052
Australia
Tel 3472526
Hon Secretary G G Brown

Bismuth Institute
Rue Brederode 9
B 1000 Brussels
Belgium
Tel 02 5132913
Telex 26876

The British Non-Ferrous Metals Federation
Crest House
7 Highfield Road
Edgbaston
Birmingham B15 3ED
United Kingdom
Tel 021 454 7766
Telex 339161
Cables Brinonfer Birmingham
Secretary R A Felton

Bureau International de la Recuperation (BIR)
Place du Samedi 13- Bte 4
1000 Brussels
Belgium
Tel 2 2178251

Centre d'Information et de Statistiques du Ferrosilicium
20 Avenue de la Gare
1003 Lausanne
Switzerland
Tel 021 238832

Chambre Syndicate du Zinc et du Cadmium
30 Avenue du Messine
F 75008 Paris
France
Tel 1 5630266/5635227
Secretary J Pommel

CIPEC (Intergovernmental Council of Copper Exporting Countries)
177 Avenue de Roule
92200 Neuilly sur Seine
Paris
France
Tel 1 7581155

Commodities Research Unit
26 Red Lion Square
London WC1 4RL
United Kingdom
Tel 01 242 7463

Copper Development Association
Orchard House
Mutton Lane
Potters Bar
Herts EN6 3AP
United Kingdom
Tel 01 775 0711

Council of Mining and Metallurgical Institutions
44 Portland Place
London W1N 4BR
United Kingdom
Tel 01 580 3802
Telex 261410
Cables Mingem
Secretary M J Jones

European Primary Aluminium Association (EPAA)
Konigsallee 30
PO Box 1207
D 4000 Dusseldorf 1
West Germany
Tel 0211 320821
Telex 8587407 alz d
Cables Aluzentrale
Secretary General Dr L Ernst

European Zinc Institute
Avenue de Messine 30
75008 Paris
France
Tel 1 33 15630266
Telex 650438

European Zinc Institute
Secretariat
PO Box 2126
5600 Eindhoven
Netherlands
Tel 31 40122497
Telex 51860

Ferro Alloy Producers' Association
PO Box 1338
Johannesburg 2000
South Africa
Tel 11 8336033
Telex 87746
Cables Vital Johannesburg
Secretary O H Hodgkin

The Ferro Alloys Association
1612 K Street NW
Washington DC 20006
USA
Tel 202 6594131

General Services Administration
General Services Building
18th and F St NW
Washington DC 20405
USA
Tel 202 6554000

International Lead/Zinc Study Group
Metro House
58 St James's Street
London SW1A 1LD
United Kingdom
Tel 01 499 9373

International Magnesium Association
1406 Third National Building
Dayton
Ohio 45402
USA
Tel 513 2230419

International Precious Metals Institute (IPMI)
Polytechnic Institute of New York
333 Jay Street
Brooklyn
New York 11201
USA
Tel 212 6433073/215 8661211

International Primary Aluminium Institute
9th Floor
New Zealand House
Haymarket
London SW1
United Kingdom
Tel 01 930 3051

International Tin Council
1 Oxendon Street
London SW1Y 4EQ
United Kingdom
Tel 01 930 0321

Japan Society of Newer Metals
Uchisaiwai Building
1-4-2 Uchisaiwaicho
Chiyoda-ku
Tokyo 100
Japan
Tel 03 5910389
Managing Director Kazuo Kuroda
Secretary Ruiko Mitsuhashi

Japan Titanium Association
Konwa Kaikan Building
12-22 1-chome Tsukiji
Chuo-ku
Tokyo
Japan
Tel 03 5424088
Cables Tsukiji Titakon
Secretary H Kodama

The Manganese Centre
17 Avenue Hoche
F 75008 Paris
France
Tel 1 5630634
Telex 641070
Director Paul L Dancoisne

The Metals Society
1 Carlton House Terrace
London SW1Y 5DB
United Kingdom
Tel 01 839 4071
Telex 8814813
Cables Themetsoc London SW1
Secretary General M J Hall

Mining & Metallurgical Society of America
Room 1352
230 Park Avenue
New York
New York 10017
USA
Tel 212 6447638
Secretary Norman H Donald Jr

Minor Metals Traders' Association
69 Cannon Street
London EC4N 5AB
United Kingdom
Tel 01 248 4444
Telex 888941
Cables Convention London
Secretary C S Algar

National Association of Recycling Industries
Inc (NARI)
330 Madison Avenue
New York NY 10017
USA
Tel 212 8677336

Non-Ferrous Metal Industries Association of
South Africa
Metal Industries House (1st Floor)
Cor Marshall & Simmonds Streets
Johannesburg 2001
South Africa
Tel 11 8336033
Telex 87746
Cables Vital Johannesburg
Secretary D A Carson

Non-Ferrous Metals Information Centre
Boulevard de Berlaimont
1000 Brussels
Belgium
Tel 02 2183504
Telex 22077

Refractory Metals Association
PO Box 2054
Princeton
New Jersey 08540
USA
Tel 609 7993300
Telex TWX 5106852516
Executive Director Kempton H Roll

Roskill Information Services Ltd
2 Clapham Road
London SW9 0JA
United Kingdom
Tel 01 582 5155

Selenium Tellurium Development
Association Inc
PO Box 3096
Darien
Connecticut 06820
USA
Tel 203 6550470
Secretary Prescott Fuller

Tantalum Producers International Study
Center
Rue aux Laines
1000 Brussels
Belgium
Tel 02 5118396/02 5125442
Cables Tictan Brussels
Secretary J A Wickens

The World Bureau of Metal Statistics
41 Doughty Street
London WC1N 2LF
United Kingdom
Tel 01 405 2771
Telex 298970
Cables Worldmetal London
Secretary J L T Davies

Zinc, Lead and Cadmium Development
Associations
34 Berkeley Square
London W1L 6AJ
United Kingdom
Tel 01 499 6636

Members of the Minor Metals Traders' Association

ACLI International
ACLI Metal & Ore Division
110 Wall Street
New York
NY 10005
USA
Tel 212 9438700
Telex RCA 232261

Ametalco Trading Ltd
29 Gresham Street
London EC2V 7DA
United Kingdom
Tel 01 606 8800
Telex 885541

Ayrton & Partners Ltd
Friendly House
21-24 Chiswell Street
London EC1Y 4SN
United Kingdom
Tel 01 638 5588
Telex 887648

Basmont Metal Co Ltd
Victoria House
Vernon Place
Southampton Row
London WC1B 4DN
United Kingdom
Tel 01 405 5065
Telex 27312

Bocmin Metals Ltd
Broadway Chambers
Hammersmith Broadway
London W6
United Kingdom
Tel 01 741 0661
Telex 934849

Brandeis Goldschmidt & Co Ltd
4 Fore Street
London EC2P 2NU
United Kingdom
Tel 01 638 5877
Telex 84401

Brookside Metal BV
Vredenberg 1.24
3511 BB Utrecht
Netherlands
Tel 30 313851
Telex 40555

Cambridge Metals Ltd
7 All Saints Passage
Cambridge CB2 3LS
United Kingdom
Tel 0223 312111
Telex 817570

Chloride Metals Ltd
Manor Metal Works
Harrow Manorway
Abbey Wood
London SE2 9RW
United Kingdom
Tel 01 310 4444
Telex 896948

Cominco (UK) Ltd
50 Finsbury Square
London EC2A 1BD
United Kingdom
Tel 01 638 4000
Telex 886563

Commodity Analysis Ltd
37-39 St Andrews Hill
London EC4V 5DD
United Kingdom
Tel 01 248 9571
Telex 883356

Continental Metals Corporation
820 Second Avenue
New York
NY 10011
USA
Tel 212 4219811
Telex 425589

Copalco International Ltd
Suite 354
200 Park Avenue
New York
NY 10017
USA
Tel 212 6977260
Telex 127758

R L Cusick (Metals) Ltd
Cusick House
Church Street
Ware
Herts
United Kingdom
Tel 0920 61181
Telex 817395

Entores Ltd
79-83 Chiswell Street
London EC1Y 4TB
United Kingdom
Tel 01 606 6050
Telex 261932

Erlanger & Co (Commodities) Ltd
Moor House
London Wall
London EC2Y 5ET
United Kingdom
Tel 01 638 6691
Telex 8813155

Exsud Ltd
237-247 Tottenham Court Road
London W1P 0BU
United Kingdom
Tel 01 631 4959
Telex 264751

FLT & Metals Ltd
1-5 Long Lane
London EC1A 9HA
United Kingdom
Tel 01 606 1272
Telex 8811917

Gerald Metals Ltd
Europe House
World Trade Centre
St Katherine-by-the-Tower
London E1 9AA
United Kingdom
Tel 01 481 0681
Telex 884377

Gill & Duffus Ltd
201 Borough High Street
London SE1 1HW
United Kingdom
Tel 01 407 7050
Telex 887588

Greendown Trading BV
Bozembocht 23
PO Box 3048
Rotterdam
Netherlands
Tel 10 13207
Telex 23508

Grondmet
Rochussenstraat 125
Rotterdam
Netherlands
Tel 10 361933
Telex 21689/25218

Intsel Ltd
83-87 Gracechurch Street
London EC3V 0AA
United Kingdom
Tel 01 623 3691
Telex 8811981

AC Israel Woodhouse Co Ltd
21 Mincing Lane
London EC3R 7DN
United Kingdom
Tel 01 623 3131
Telex 8831136/883139

Lambert Metals Ltd
506-508 Kingsbury Road
London NW9 9HE
United Kingdom
Tel 01 204 9422
Telex 23844

Leopold Lazarus Ltd
Gotch House
20-34 St Bride Street
London EC4A 4DL
United Kingdom
Tel 01 583 8060
Telex 265544

Leigh & Sillavan Ltd
Knights Pool
Windmill Street
Macclesfield
Cheshire SK11 7HR
United Kingdom
Tel 0625 31331
Telex 668363

Lonconex Ltd
29 Mincing Lane
London EC3R 7EU
United Kingdom
Tel 01 626 4383
Telex 885016

Maclaine Watson & Co Ltd
2-4 Idol Lane
London EC3R 5DL
United Kingdom
Tel 01 283 8611
Telex 883854

Arthur Matyas & Co Ltd
Pearl House
746 Finchley Road
London NW11 7TH
United Kingdom
Tel 01 458 8911
Telex 8812123

Metaleg Metall GmbH
Graf Adolph Strasse 22
4000 Dusseldorf 1
West Germany
Tel 211 631011
Telex 8582610

Metallbodio Ltd
PO Box 296
CH-4010 Basle
Switzerland
Tel 61 238953
Telex 62270

Metramet Ltd
Kingswell
58-62 Heath Street
London NW3 1EN
United Kingdom
Tel 01 794 1131
Telex 25479

Minor Metals Inc
1 Gulf & Western Plaza
Room 918
New York NY 10023
USA
Tel 212 5418880
Telex 426965

Minwood Metals Ltd
Paterson Road
Finedon Industrial Estate
Wellingborough
Northants
United Kingdom
Tel 0933 225766
Telex 311394

Nemco Metal International Ltd
9 Harrowden Road
Brackmills
Northampton NN4 0EB
United Kingdom
Tel 0604 66181
Telex 826433

A J Oster Co
50 Sims Avenue
Providence
Rhode Island
USA
Tel 401 4213840
Telex WUD 927747

Powell Metals & Chemicals (UK) Inc
62 Hills Road
Cambridge CB2 1LA
United Kingdom
Tel 0223 51775
Telex 817669 Powcang

Primetal Italia
20124 Milano
Via Cappellini 16
Milan
Italy
Tel 2 667124
Telex 33439

Derek Raphael & Co Ltd
DRC House
2 Cornwall Terrace
Regents Park
London NW1 4QP
United Kingdom
Tel 01 486 9931
Telex 261916

Redlac Metals Ltd
148 Buckingham Palace Road
London SW1W 9TR
United Kingdom
Tel 01 703 2276
Telex 888885

Rhondda Metal Co Ltd
Rhondda Works
Perth
Mid-Glamorgan CF39 9BA
United Kingdom
Tel 044 3612881
Telex 49562

Rothmetal Trading Ltd
PO Box 1549
Argus Insurance Building
Wesley Street
Hamilton
Bermuda
Tel 80929 27980
Telex 3368 BA

William Rowland Ltd
Powke Lane
Cradley Heath
Warley B64 5PX
United Kingdom
Tel 021 559 3031
Telex 331376

Sassoon & Co SA
203 Avenue Louise
PO Box 4
1050 Brussels
Belgium
Tel 2 6406783
Telex 63605

Jack Sharkey & Co Ltd
Middlemore Road
Smethwick
Warley B66 2DP
United Kingdom
Tel 021 558 7444

Skandinaviska Malm-Och
Metallaktiebolaget
Kungsgatan 6
Box 7547
S-103 93 Stockholm
Sweden
Tel 8 233520
Telex 19552

Sogemet
161 Avenue Charles de Gaulle
92202 Neuilly sur Seine
Paris
France
Tel 1 6375760
Telex 820242

Spencer Metals & Minerals Ltd
25 London Road
Newbury
Berks
United Kingdom

Steetley Chemicals Ltd
Berk House
PO Box 56
Basing View
Basingstoke
Hants RG21 2EG
United Kingdom
Tel 0256 29292
Telex 858371

Sterling Enterprises Metals Ltd
Sterling House
328 Holloway Road
London N7 7HJ
United Kingdom
Tel 01 607 7381
Telex 27325

Strategic Metals Corporation
500 Chesham House
150 Regent Street
London W1R 5FA
United Kingdom
Tel 01 439 6288

Tennant Trading Ltd
9 Harp Lane
Lower Thames Street
London EC3R 6DR
United Kingdom
Tel 01 626 4533
Telex 884724

Trans-World Metals Ltd
Walsingham House
35 Seething Lane
London EC3N 4EL
United Kingdom
Tel 01 480 5701
Telex 8951322

Unimet GmbH
Stahl, Rohre und Metalle
Cecilienallee 21
4000 Dusseldorf 30
West Germany
Tel 211 450914
Telex 08582622

H A Watson & Co Ltd
119-120 High Street
Stourbridge DY8 1DT
United Kingdom
Tel 03843 77801
Telex 337034

Rene Weil SA
77 Rue de Monceau
75008 Paris
France
Tel 1 5630488
Telex 280445

Wheatstock Ltd
4 Fore Street
London EC2Y 5EH
United Kingdom
Tel 01 588 7081
Telex 8814900

Wogen Resources Ltd
17 Devonshire Street
London W1
United Kingdom
Tel 01 580 5762
Telex 28820

Journals

American Metal Market
7 East 12th Street
New York NY 10003
USA
Tel 212 7414000

Metal Bulletin Ltd
45-46 Lower Marsh
London SE1 7RG
United Kingdom
Tel 01 633 0525

Metal Bulletin Inc
708 3rd Avenue
New York NY 10017
USA
Tel 212 4900791

The Metals Investor
711 West 17th Street, G-6
Costa Mesa CA 92627
USA
Tel 714 6420243

Metals Week
McGraw-Hill Inc
1221 Avenue of the Americas
New York NY 10020
USA
Tel 212 9971221

Tin International
Tin Publications Ltd
7 High Road
London W4 2NE
United Kingdom
Tel 01 995 9277

Appendix 2:
Glossary of relevant terms

Actuals: Physical commodities, also commodities readily available. The commodity itself as opposed to a futures contract

Agent: A merchant who acts on behalf of a producer usually on a commission basis

Anode: The electrode at which negatively charged ions are discharged during electrolysis

Arbitrage: Purchase of contract in one market while simultaneously selling the same amount in another to take advantage of price differentials

Backwardation: Market description of the situation when the spot or nearby prices are higher than those for future delivery months. Usually caused by delays in shipment thus creating shortages in available supplies. Opposite of contango

Basic price: Agreed price between buyer and seller of an option at which the option may be taken up. Also called the 'striking price'

Bear: Person expecting a decline in prices

Bear covering: Closing of short positions

Bid: The price which the buyer is willing to pay

Borrowing: Purchase of a nearby delivery date and simultaneous sale of a forward date. Used only in London Metal Exchange

Brine: Solution of any salt

Broker: Establishes contact between buyer and seller, for a fee. In the US, ring dealing members of futures markets are frequently called brokers

Bulk: Delivery of unpacked metal

Bull: Person expecting a rise in prices

Call: A period for trading. Conducted by a chairman to establish a price for a specific time. During a call, trading is confined to one delivery month

Call option: The option buyer/taker pays a premium and holds the right to decide at a later stage whether or not to buy at the price agreed at the time the premium was paid. The right may be exercised at any time from the point of purchase to the expiry of the option

Carrying: General term covering both *Borrowing* and *Lending*

Carrying Costs: Costs connected with warehouse storage, insurance, etc. On occasion includes interest and estimated changes in weight

Casting: Metal object made by pouring molten metal into a mould, which does not involve mechanical work such as rolling or forging

Catalyst: A chemical (metal) which assists a chemical reaction but remains chemically unchanged at the end

Cathode: The electrode at which positively charged (metallic) ions are discharged during electrolysis, hence the slab of metal deposited at the end of the electrolytic process

Certified stocks: Supplies rates as deliverable

C&F: Cost and freight

CIF: Cost, insurance and freight (included in price)

Clearing house: The organization that provides clearing facilities for some futures markets

Commission house: A company, which trades on behalf of clients for a commission. The Commission House only handles clients' business and does not trade on its own account

Concentrate: The ore of a metal after separation from other unwanted minerals

Conductivity: A measure of the ability of a metal to conduct an electric current

Contango: A situation where prices are higher in the forward delivery months than in the nearby delivery month. Opposite of backwardation. Normally in evidence when supplies are adequate or in surplus. The contango reflects either wholly or in part the costs of holding and financing

Contract: An agreement to buy or sell a specified amount of a particular commodity. It details the amount and grade of the product and the date on which the contract will mature and become deliverable, if not previously liquidated

Contract month: Month in which a given contract becomes deliverable, if not liquidated or traded out before the date specified

Custom smelter: A smelter which relies on concentrate purchased from independent mines instead of its own captive sources

Delivery basis: Specified locations to which the commodity in a futures contract may be physically delivered in order to terminate the contract

Delivery date: Or Prompt Date, on which the commodity must be delivered to fulfil the terms of the contract

Delivery month: Calendar month stipulated as month of delivery in a futures contract

Deposit: Sum of money required by the broker from his client, usually 10 per cent of the value of the contract, to justify opening of a futures position

Differentials: Premiums paid for grades better than the basic grade, or discounts allowed for grades below the basic grade

Double option: This is an option which gives the buyer or person taking the option the right either to buy from or sell to the seller of the option or the person who gives it, at the strike price

Ductility: Quality of a metal by which it may be beaten into sheet or drawn into a wire

Electrolysis: A process used for refining metals where the metal is deposited on a cathode from a solution or molten mass

Element: A substance which cannot be split into anything simpler in a chemical process

Fabricator: A company which makes semi-fabricated products from refined metal and on occasions from scrap

Fixation: Fixing price in the future, and used in commodity call purchases and call sale trades

Flotation: Process by which ore is separated from earth and rubble, during which the ore is introduced to a bath of liquid which is agitated so that the metallic ore either rises to the top or sinks to the bottom

Flux: A substance added to a metal in order to make it easier to melt

FOB: Free on Board

Force Majeure: This is a clause in a supply contract which permits either party not to fulfil the contractual commitments due to events beyond their control. These events may range from strikes to export delays in producing countries

Forward shipment: Contract covering actual commodity shipments at a specified date in the future

Futures contract: Contract which requires the delivery of a commodity in a specified future month, if not liquidated before the contract matures

Grades: Standards set for judging the quality of a commodity

Hedge: A temporary futures market sale which is made against a spot purchase, or alternatively a temporary futures market purchase made against a spot sale. The purpose is to reduce risk from price fluctuations on the physical transaction until the reverse futures market operation cancels the hedge, or liquidates the original operation

Inorganic: Compounds other than those of carbon, but including the oxides of carbon and carbonates

Integrated producer: A producer who owns mines, smelters and refineries and also, in some instances, fabricating plants

Kerb trading: Unofficial trading when the market has closed. The term 'kerb' dates from the time when dealers continued trading on the kerb outside the exchanges after these had closed

Last trading day: Final day for trading a particular delivery. Positions which have not been closed by the last trading day must be fulfilled by making, or taking, delivery of the physical commodity or metal

Leaching: The process by which metals are extracted from a low grade ore or waste product, by the direct action of a caustic solution or an acid, producing a metallic salt solution

Lending: Sale of a nearby delivery date coupled with the simultaneous purchase of a more distant date (LME term)

Limit up/down: Largest permitted rise/fall in the price of a commodity traded in a futures contract during a trading session, as fixed by the contract market's rules

Liquidation: Sale of long contract to offset previous purchase. Operation which cancels an earlier position

Long: An open purchased futures contract. Buying forward on the market

Lot: Minimum contract unit in a hedge or futures market

Margin: This is the amount deposited as a guarantee for the fluctuations on a futures purchase or sale. If the contract fluctuates against the holder of the contract, he is required to provide for the difference between his contract price and the current market price by paying 'variation margin' differences. Thus the original margin continues to guarantee fully the performance of the contract at any market price level

Margin call: A commodity broker's request to a client for additional funds to secure the original deposits

Master alloy: An alloy made with two metals x and y, which is then added to a larger quantity of metal x so that metal y is diluted to the required proportion

Nearby delivery: The nearest active month of delivery on a futures market

Offer: The seller's price for the commodity offered

Open contracts: Contracts bought or sold and not offset by an opposite trade

Open outcry: Trading conducted by calling out bids and offers across a ring or pit and having them accepted

Open position: A forward market position which has not been closed out

Option: The holder of the option has the right to buy from or sell to the granter of the option a specified quantity of the commodity at an agreed price. The cost of buying the option is known as the premium

Pegged price: The price at which a commodity has been fixed by agreement

Plating: The deposition of a layer of a metal on an object forming the cathode during electrolysis

Position trader: Someone who takes long or short positions in futures markets in consequence of an opinion that prices are about to advance or decline

Premium: The amount by which a cash commodity price sells over a futures price or another cash commodity price. The excess of one futures contract price over another

Put option: This option gives the buyer (or 'taker') of the option – in exchange for the premium which he pays – the right to decide at a later date whether or not to sell to the seller (or 'granter') at the price ('basic' or 'striking' price) agreed at the time the premium was paid. The right may be exercised at any point from the purchase of the option up to the declaration date (the date upon which the option expires)

Reduction: A process involving the removal of oxygen or a group which forms negative ions

Ring: Space on a trading floor where futures are traded. Also known as a pit in the US

Roast: Heating ore or concentrate in air or inert atmosphere to extract unwanted element, usually sulphur

Salt: A compound containing a metal and a non-metal

Scavenger: Substance added to molten metal which has the effect of ridding the melt of an unwanted element

Semi-fabrication: The production of metal in the form of wire, sheet, rod etc

Short: The sale of a commodity not owned by the seller

Slime: Substance containing metal other than that deposited at the cathode during electrolysis and which is found as a sludge in the bottom of the electrolytic cell

Smelt: Extraction of crude metal from the ore by melting, prior to refining

Sponge: Lumpy form of metal with sponge-like appearance produced by casting molten metal into water

Spot: Term denoting immediate delivery for cash, as distinct from future delivery

Spot month: The first month in which delivery can take place and for which a quotation is made on the futures market

Spot price: The commodity cash sale price, as opposed to a futures price

Spread: An order to purchase one contract month and sell another month in the same commodity, usually done on the same exchange

Squeeze: Pressure on a delivery date which results in the price of that date becoming firmer in relation to other dates

Straddle: The simultaneous buying and selling of the same commodity on the same market. This is designed to take advantage of differences between two options. Example – the sale of a September option and the simultaneous purchase of a January option made in the expectation that a later simultaneous purchase of the September and sale of the January options will produce a profit. See *Arbitrage*

Sublimation: Heating an unrefined metal until it turns into its gaseous form. This is then condensed into the refined metal

Switch: To advance or postpone the original contract to a different month

Switching: Exchanging a commodity in one warehouse for a commodity in another

Terminal market: Usually synonymous with commodity exchange or futures market, especially in the United Kingdom

Unwrought: Metal in a cast condition, unworked by mechanical means

Warehouse receipt: A receipt for a commodity given by a licensed or authorized warehouseman and issued as tender on futures contracts

Warrant or Warehouse receipt: A receipt of physical deposit which gives title to the physical commodity

Appendix 3:

Average monthly dollar/sterling rate based on the market price at 11.30 am each dealing day

	January	February	March	April	May	June	July	August	September	October	November	December	Average
1958	2.8000	2.8000	2.8000	2.8000	2.8000	2.7999	2.7979	2.7978	2.7981	2.7980	2.7995	2.7994	2.7992
1959	2.8047	2.8077	2.8103	2.8148	2.8127	2.8109	2.8101	2.8080	2.8020	2.8050	2.8012	2.7970	2.8070
1960	2.7985	2.8017	2.8041	2.8078	2.8046	2.8010	2.8065	2.8084	2.8107	2.8096	2.8119	2.8057	2.8059
1961	2.8047	2.7986	2.7956	2.7965	2.7922	2.7888	2.7854	2.8006	2.8096	2.8136	2.8130	2.8078	2.8005
1962	2.8089	2.8126	2.8132	2.8122	2.8105	2.8068	2.8050	2.8023	2.7995	2.7999	2.8001	2.8019	2.8061
1963	2.8031	2.8018	2.7990	2.7990	2.7980	2.7986	2.7993	2.7981	2.7961	2.7964	2.7964	2.7951	2.7984
1964	2.7967	2.7955	2.7966	2.7974	2.7979	2.7926	2.7884	2.7852	2.7819	2.7820	2.7833	2.7887	2.7905
1965	2.7895	2.7932	2.7904	2.7938	2.7953	2.7904	2.7892	2.7891	2.7945	2.8014	2.8020	2.8003	2.7941
1966	2.8025	2.8010	2.7938	2.7918	2.7908	2.7884	2.7873	2.7875	2.7877	2.7901	2.7896	2.7884	2.7916
1967	2.7892	2.7924	2.7943	2.7974	2.7952	2.7899	2.7860	2.7838	2.7829	2.7817	2.6430	2.4062	2.7452
1968	2.4087	2.4094	2.3992	2.4015	2.3891	2.3845	2.3892	2.3912	2.3869	2.3895	2.3874	2.3838	2.3934
1969	2.3866	2.3910	2.3913	2.3928	2.3862	2.3896	2.3902	2.3855	2.3837	2.3896	2.3962	2.3967	2.3900
1970	2.3999	2.4042	2.4056	2.4059	2.4033	2.3976	2.3903	2.3877	2.3845	2.3870	2.3899	2.3903	2.3955
1971	2.4047	2.4170	2.4182	2.4175	2.4183	2.4179	2.4182	2.4342	2.4709	2.4915	2.4937	2.5267	2.4441
1972	2.5703	2.6030	2.6178	2.6093	2.6117	2.5814	2.4415	2.4502	2.4425	2.3968	2.3518	2.3456	2.5018
1973	2.3558	2.4418	2.4778	2.4825	2.5174	2.5743	2.5402	2.4742	2.4163	2.4260	2.3890	2.3171	2.4510
1974	2.2263	2.2738	2.3369	2.3906	2.4117	2.3876	2.3869	2.3445	2.3146	2.3310	2.3271	2.3271	2.3382
1975	2.3603	2.3917	2.4172	2.3680	2.3185	2.2807	2.1833	2.1133	2.0852	2.0548	2.0499	2.0208	2.2203
1976	2.0276	2.0257	1.9455	1.8476	1.8156	1.7710	1.7863	1.7819	1.7309	1.6401	1.6370	1.6787	1.8073
1977	1.7127	1.7100	1.7172	1.7195	1.7188	1.7193	1.7221	1.7403	1.7436	1.7703	1.8203	1.8550	1.7458
1978	1.9338	1.9405	1.9103	1.8507	1.8175	1.8366	1.8940	1.9419	1.9593	1.9740	1.9905	1.9950	1.9203
1979	2.0030	2.0040	2.0351	2.0759	2.0541	2.1029	2.2495	2.2326	2.2005	2.1472	2.1308	2.1966	2.1190
1980	2.2676	2.2908	2.1982	2.2146	2.2913	2.3460	2.3570	2.3490	2.3970	2.3880	2.4370	2.3585	2.3246
1981	2.3920	2.3680	2.2190	2.2450	2.1410	2.0705	1.9805	1.8060	1.8360	1.8510			

Appendix 4:
World Time Zones

Standard Times at noon Greenwich Mean Time

British Summer Time, which is one hour ahead of GMT, is observed from 02.00 hours on the fourth Sunday in March until 02.00 on the fourth Sunday in October. USA Daylight Saving Time, which is one hour ahead of local standard time, is observed in all states except Arizona, Hawaii and Michigan from 02.00 on the last Sunday in April until 02.00 on the last Sunday in October.

City	Time	City	Time	City	Time
Accra	12.00	Damascus	14.00	Ottawa	07.00
Adelaide	21.30	Darwin	21.30	Panama	07.00
Algiers	13.00	Delhi	17.30	Paris	13.00
Amman	14.00	Djakarta	20.00	Peking	20.00
Amsterdam	13.00	Dublin	12.00	Perth	20.00
Ankara	14.00	Gibraltar	13.00	Prague	13.00
Athens	14.00	Helsinki	14.00	Quebec	07.00
Auckland	24.00	Hobart	22.00	Rangoon	18.30
Baghdad	15.00	Hong Kong	20.00	Rawalpindi	17.00
Bangkok	19.00	Istanbul	14.00	Reykjavik	12.00
Beirut	14.00	Jerusalem	14.00	Rio de Janeiro	09.00
Belgrade	13.00	Karachi	17.00	Rome	13.00
Berlin	13.00	Kuala Lumpur	20.00	San Francisco	04.00
Berne	13.00	Lagos	13.00	Santiago	08.00
Bombay	17.30	Leningrad	15.00	Singapore	19.30
Bonn	13.00	Lima	07.00	Sofia	14.00
Brisbane	22.00	Lisbon	13.00	Stockholm	13.00
Brussels	13.00	Luxembourg	13.00	Sydney	22.00
Bucharest	14.00	Madras	17.30	Tehran	15.30
Budapest	13.00	Madrid	13.00	Tokyo	21.00
Buenos Aires	09.00	Melbourne	22.00	Toronto	07.00
Cairo	14.00	Mexico City	06.00	Tunis	13.00
Calcutta	17.30	Montevideo	08.30	Vancouver	04.00
Canberra	22.00	Moscow	15.00	Vienna	13.00
Cape Town	14.00	Nairobi	15.00	Warsaw	13.00
Caracas	08.00	New York	07.00	Washington	07.00
Chicago	06.00	Nicosia	14.00	Wellington	24.00
Copenhagen	13.00	Oslo	13.00	Winnipeg	06.00

Note to World Time Zones map
The earth turns one complete revolution in 24 hours. The surface of the earth is divided into 24 Time Zones, each of 15° longitude or 1 hour of time. In 24 hours it turns through 360°. The times shown are the standard times on land and sea when it is 12.00 hours on the Greenwich Meridian.

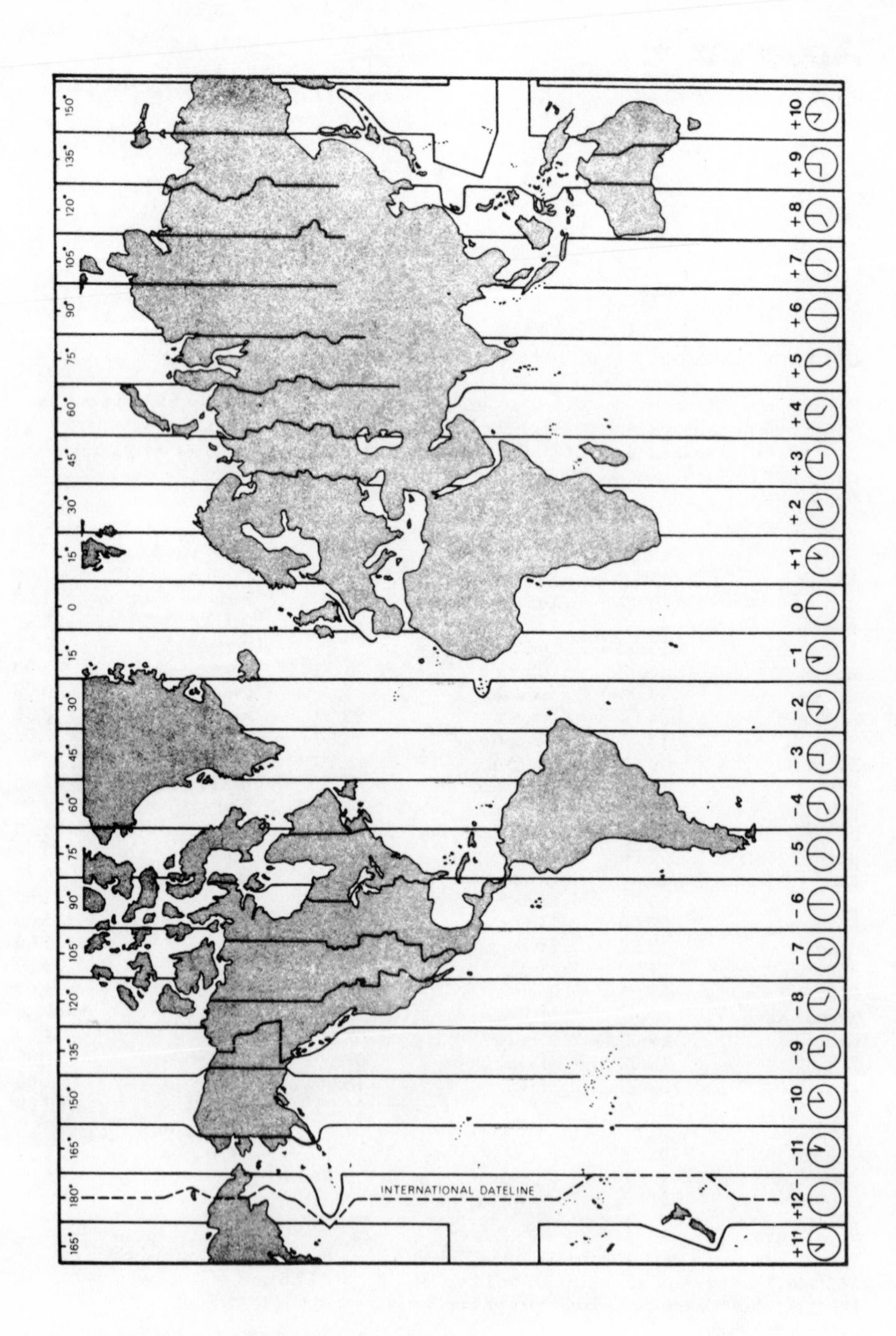
150° 135° 120° 105° 90° 75° 60° 45° 30° 15° 0 15° 30° 45° 60° 75° 90° 105° 120° 135° 150° 165° 180° 165°
+10 +9 +8 +7 +6 +5 +4 +3 +2 +1 0 −1 −2 −3 −4 −5 −6 −7 −8 −9 −10 −11 +12− +11
INTERNATIONAL DATELINE

Appendix 5:

London Metal Exchange contract rules and arbitration

CONTRACTS

Rule A

Members of the London Metal Exchange, in their dealings with other Members, shall be responsible to and entitled to claim against one another, and one another only, for the fulfilment of every Contract for Metals.

Rule B

In these Rules the expression "Members of the London Metal Exchange" includes Firms and Companies who, although not themselves Subscribers to the Exchange, are represented and deal thereon by and through "Representative Subscribers" to the Exchange acting as the representatives or Agents of such Firms or Companies.

Rule C

If any Member of the Metal Exchange fails to meet his engagements to another Member, whether by failing to provide on the due date documents (i.e. Bills of Lading, Warrants or Delivery Orders according to the metals dealt in) to meet sales made or money to pay for metals bought, or by making default in fulfilling any other obligation arising out of dealings made subject to the Rules and Regulations of the London Metal Exchange, notice of the default shall be given at once in writing to the Committee of the Exchange and the Committee shall immediately fix and publish a settlement price or prices as at the date of such communication to them for all contracts which the defaulter may have open under these Rules, whether with Members or with parties who are not Members. All such contracts shall forthwith be closed and balanced, by selling to or buying from the defaulting Member such metals as he may have contracted to deliver or take, at the settlement prices fixed for this purpose by the Committee, and any difference arising whether from or to the party in default shall become payable forthwith notwithstanding that the prompt day or other day originally stipulated for the settlement of the transaction may not have arrived. In fixing settlement prices under this Rule the Committee may in their discretion take into consideration the extent and nature of the transactions which the defaulting Member has open and any other circumstance which they may consider should affect their decision. In any case where the Committee shall be of opinion that the default is not due to the insolvency of the defaulter the Committee shall by resolution negative the application of this rule. Any claim arising out of a default not due to insolvency shall be settled by arbitration in the usual manner. This rule shall apply to cases in which at or after the decease of a Member the engagements entered into by him are not duly met.

Rule D

In any Contract made subject to the Rules and Regulations of the London Metal Exchange between a Member and a Non-Member in the event of the Non-Member failing to meet his engagements arising out of any such contract whether by failing to provide on the due date documents to meet sales or money to take up documents (as the case may be) or otherwise howsoever or of his failing to supply or maintain such margin (if any) for which the Member is entitled to call and has called, or in the event of the Non-Member's suspending payment or becoming bankrupt or committing any act of bankruptcy or (being a Company) in the event of its going into liquidation whether voluntary or otherwise, the Member shall have the right to close all or any such Contracts outstanding between them by selling out or buying in against the Non-Member (as the case may be) and any differences arising therefrom shall be payable forthwith notwithstanding that the prompt day or other day originally stipulated for settlement may not have arrived.

Rule E

Payments for Warrants or other documents (when deliverable under the Contracts) unless otherwise stipulated on the Contract, shall be made by cash in London, or by cheque on a London clearing bank, either mode in Seller's option. The documents shall be tendered in London against the cash or cheque, as the case may be, and not later than 2.30 p.m. on the prompt or settling day.

Rule F

Contracts wherein Buyer or Seller (as the case may be) has the option to uplift or to deliver, prior to the prompt or settlement date by giving previous notice of his intention, shall have the notice reckoned by market days; such notices, unless otherwise stipulated at time of purchase or sale, shall be as follows: On a Contract with the option to uplift or to deliver during one calendar month or less, one day's notice shall be given; on a Contract with the option beyond one and up to two calendar months two days' notice shall be given; and on a Contract with the option beyond two and up to three calendar months three days' notice shall be given previous to the date on which delivery is required, or will be made. Notice shall be given for the whole quantity stated in the contract and shall be tendered in writing and delivered at the office of the Seller of the option not later than noon on the day of notice. Rent shall only be allowed to Buyer to actual day of settlement; and there shall not be any allowance of interest for a payment made prior to the prompt date.

Rule G

Prompt or settlement dates falling on Saturday, Sunday, or a Bank Holiday, which days are not market days, shall be settled as follows: Prompts falling on Saturday shall be settled on the Friday previous; but should the preceding Friday be a Bank Holiday the prompt shall be extended to the Monday following; should both the Friday preceding and the Monday following be Bank Holidays, the prompt shall be settled on the Thursday previous. Prompts falling on Sunday should be extended to the Monday following, but should that Monday be a Bank Holiday the prompt shall be extended to the Tuesday following; should both the following Monday and Tuesday be Bank Holidays, the prompt shall then be extended to the Wednesday following. Prompts falling on a Bank Holiday shall be extended to the day following; and if the Bank Holiday falls on Friday the prompt shall be extended to the Monday following; but should the Friday be Good Friday, prompts falling on that day shall be settled on the Thursday previous. If Christmas Day falls on Monday, prompts falling on that day shall be extended to the Wednesday following, but if Christmas Day falls on Tuesday, Wednesday, Thursday or Friday, prompts falling on that day shall be settled on the day previous.

Rule H
The establishment, or attempted establishment of a "corner", or participation directly or indirectly in either, being detrimental to the interests of the Exchange, the Committee shall, if in their opinion a "corner" has been or is in the course of being established, have power to investigate the matter and to take whatever action it considers proper to restore equilibrium between supply and demand. Any member or members may be required to give such information as is in his or their possession relative to the matter under investigation.

Rule J (OPTIONS)
On the day on which notice is due, the holder of the option shall, except in cases to which Rule C applies, declare in writing before 11.30 a.m. whether he exercises or abandons the option, and if he fails to make such declaration the option shall be considered as abandoned. Options (subject to Rule F above) may be declared for less than the total optional quantity in quantities of 25 tonnes for Copper-Electrolytic Wirebars, H.C.F.R. Wirebars, Cathodes or Fire Refined, 5 tonnes for Standard Tin, High Grade Tin, 25 tonnes for Aluminium, 10,000 troy ounces for Silver, and 6 tonnes for Nickel, or multiples thereof, only one declaration against each contract being allowed. In cases to which Rule C applies the prices fixed by the Committee, at which outstanding contracts are to be closed, shall equally apply to all option contracts; and all options shall be automatically determined, and be deemed to have been either exercised or abandoned according as the prices may be in favour of or against the defaulter and whether the defaulter be the Seller or the Buyer of an option, and the option money shall be brought into account. In contracts with optional prompts, the price which shall be taken as the basis of settlement shall be the settlement price fixed by the Committee under Rule C for the prompt most favourable to the holder of the option.

Rule K (CLEARING)
All contracts made between Members of the London Metal Exchange who are entitled to deal in the Ring, either for Copper-Electrolytic Wirebars, H.C.F.R. Wirebars, Cathodes or Fire Refined, Standard Tin, High Grade Tin, Standard Lead, Standard Zinc, Aluminium, Nickel or Silver, shall be settled through the Clearing, except when a Member insists on his right to receive cash instead of cheque from the Member to whom he has sold, in which case the Seller shall give notice to his Buyer before noon on the market day preceding the settling day, and such transactions shall then be exempted from settlement through the Clearing. The Rules governing the Clearing of all contracts shall be those in existence at the time fixed for the fulfilment of the contract. Copies of such rules may be obtained from the Secretary of the Exchange.

Rule L
In case of strikes, lock-outs, or other unforeseen contingencies in London, or other authorised port or point of delivery, which prevent or delay the discharge and/or warehousing of Copper-Electrolytic Wirebars, H.C.F.R. Wirebars, Cathodes or Fire Refined, Standard Tin, High Grade Tin, Standard Lead, Standard Zinc, Aluminium, Nickel and/or Silver, the Seller may be allowed to postpone delivery if he can prove to the satisfaction of the Committee (of which proof the Committee shall be the sole judge) that he does not hold available metal in warehouse or vault with which to fulfil his contracts and that he has metal of the requisite quality which has arrived in London or any other authorised port or point of delivery at least ten days prior to the earliest prompt for which relief is asked, or has metal of the requisite quality in his works, but the delivery, discharge and/or other warehousing of which is prevented or delayed as aforesaid. He must also deposit with the Secretary of the

Exchange such sums as the Committee may require but not exceeding £5 per tonne in the case of Copper, Lead, Zinc and Aluminium, £10 per tonne in the case of Tin and Primary Nickel, and £5 per thousand ounces in the case of Silver. No interest will be allowed on deposits, which will be returned after delivery of Warrants. Should his application be passed by the Committee, he shall deposit documents or other proof to the satisfaction of the Committee with the Secretary of the Exchange, who shall issue Certificates for Copper, Lead, Zinc and Aluminium in quantities of 25 tonnes, Certificates for Nickel in quantities of 6 tonnes, Certificates for Tin in quantities of 5 tonnes, and Certificates for Silver in quantities of 10,000 troy ounces. The Seller shall deliver these Certificates to his Buyer. The Certificates will then constitute a good delivery on the Clearing within the period stated thereon and differences must be settled on the prompt day. The holder of a Certificate must present it to the firm named thereon not later than 2.30 p.m. on the day following that on which he receives notice in writing from his Seller that the Warrant for the actual Copper, Tin, Lead, Zinc, Aluminium, Nickel or Silver is ready. He must take up the Warrant against payment at the settlement price fixed on the preceding market day, receiving or paying any difference between this and the price mentioned on the Certificate. In the event of the price on the Certificate being above or below the settlement price operative on the day of delivery the receiver shall pay or be paid the amount of any difference. No other payments shall pass except against delivery of the actual Warrant. In case of any dispute, the Committee's ruling to be final. A fee of £5 to be paid by the Applicants for each Certificate issued.

ARBITRATION

Rule 1

All disputes arising out of or in relation to contracts subject to the Rules and Regulations of the London Metal Exchange shall be referred to arbitration as hereinafter provided. The Executive Secretary of the Committee of the London Metal Exchange (hereinafter referred to as "the Secretary") shall be notified of such disputes in writing and the party first notifying the difference shall at the time of such notification deposit with the Metal Market & Exchange Co. Ltd., the sum of £100. All such disputes shall be referred to two arbitrators, one to be appointed by each party to the difference from the Arbitration Panel of the London Metal Exchange, such arbitrators having power to appoint a third arbitrator from the Panel and having all the powers conferred on arbitrators by the Arbitration Act 1950 or any statutory modifications thereof for the time being in force. The Secretary shall be notified in writing by each party of the appointment of the arbitrators. The arbitration and any Appeal made pursuant to Rule 8 of these Rules from the Award of the Arbitrators to the Committee shall take place at the London Metal Exchange (unless mutually agreed by the Arbitrators and the parties to the dispute that the venue should be elsewhere in England or Wales) and English procedure and law shall be applied thereto.

Rule 2

Persons eligible for appointment to the Arbitration Panel shall be members of the Exchange, their partners or co-directors (as the case may be) or members of their staff. Appointment to and removal from the Panel shall be made, at their sole discretion, by the Committee of the London Metal Exchange who will also be responsible for maintaining a panel of sufficient size.

Rule 3

In the event of either party to the difference (a) failing to appoint an arbitrator, or (b) failing to give notice in writing or by cable of such appointment to reach the other party within 14 days after receiving written or cabled notice from such other party of the appointment of an arbitrator (any notice by either party being given to the other either by cable or by registered post addressed to the usual place of business of such other party), or (c) in the case of death, refusal to act, or incapacity of an arbitrator, then, upon written or cabled request of either party an arbitrator shall be appointed from the said Arbitration Panel by the Committee of the London Metal Exchange.

Rule 4

In case the two arbitrators appointed as aforesaid, whether originally or by way of substitution, shall not within three calendar months after the appointment of the arbitrator last appointed deliver their Award in writing, or choose a third arbitrator, then the said Committee on the written request of either party shall appoint a third arbitrator selected from the said Arbitration Panel to act with the two aforesaid arbitrators.

Rule 5

The Award in writing of the arbitrators or any two of them shall be made and delivered in triplicate to the Secretary within a period of three calendar months from the date of the acceptance of the appointment by the arbitrator last appointed.

Rule 6

Every Award made pursuant to any provision of this Rule shall be conclusive and binding on the parties to the arbitration, subject to appeal as hereinafter mentioned.

Rule 7

The procedure upon an arbitration shall be as follows:

(a) Within a period of 21 days after the appointment of the second of the two arbitrators so appointed, each party shall deliver to the arbitrators and to each other a statement of case in writing with the originals, or copies, of any documents referred to therein. All such documents to be in the English language or accompanied by certified translations into English.

(b) If either party shall make default in delivering such statements and documents (due consideration being given to time occupied by mails), the arbitrators shall proceed with the case on the statement before them, provided always that, in the sole discretion of the arbitrators, an extension of time may be allowed for the delivery of such statements and documents.

(c) The arbitrators shall appoint a day for a hearing within 28 days, or such further time as the arbitrators shall in their sole discretion allow, after the expiry of the 21 days in accordance with Rule 7 (a), and shall give due notice in writing thereof to the parties, who may, and if required by the arbitrators shall, attend and shall submit to examination by the arbitrators and produce such books and documents as the arbitrators may require. Each party shall be entitled to produce verbal evidence before the arbitrators.

(d) Neither Counsel, nor Solicitor shall be briefed to appear for either party without the consent of the arbitrators.

(e) The arbitrators may engage legal or other assistance.

(f) The arbitrators may adjourn the hearing from time to time, giving due notice in writing to the parties of the resumed hearing, and the arbitrators may, if they think fit, proceed with such a resumed hearing in the absence of either party or of both parties.

(g) Where any change takes place in the constitution of the tribunal of arbitrators, either by substitution or otherwise, the new tribunal shall appoint a day for the hearing which shall be not later than 28 days, nor earlier than 7 days, after the change. Each party, if desiring to do so, may submit an Amended Statement of Case, with a copy to the other party, which must reach the new tribunal within seven days of its appointment.

(h) In the event of a third arbitrator being appointed, the provisions contained in Section 9 Sub-Section 1 of the Arbitration Act 1950 shall not apply to any reference.

(i) The cost of the arbitration shall be at the sole discretion of the arbitrators. The arbitrators shall fix the amount of their remuneration. The Award shall state separately the amount of such costs and remuneration and by whom they shall be paid and whether the whole or any part of the deposit referred to in Rule 1 of these Rules shall be returned to the party lodging the same or be forfeited. In the event of either or both parties having been granted permission by the arbitrators to be legally represented at the hearing the arbitrators may take into consideration any legal costs which have been incurred.

(j) The Award shall be deposited with the Secretary who shall forthwith give notice of receipt thereof in writing to both parties, and a copy of such Award shall be delivered to both parties on payment by either party of the costs specified in the Award, which payment shall not affect any provision of the Award.

(k) In the event that after the deposit referred to in Rule 1 of these Rules has been made the parties to the arbitration shall (i) settle their differences (ii) fail to proceed as directed by the arbitrators under sub-clause (c) of this Rule (iii) fail to take up the Award within 28 clear days of notification being given under sub-clause (j) of this Rule, such deposit shall be forfeited.

(l) At the time of issuing their Award, all statements and all documents lodged with the arbitrators shall be delivered by them to the Secretary, by whom they shall be retained until the expiration of the time for giving notice of appeal, as hereinafter mentioned, after which the Secretary shall, unless there shall be such appeal, return them to the parties concerned.

Rule 8
Either party shall have the right to appeal against the Award to the Committee of the London Metal Exchange.

Rule 9
The method of appeal against the Award shall be as follows:

(a) The party making the appeal shall (i) within 21 days of the date of the Award give notice in writing of such appeal to the Secretary, and to the other party and shall at the same time state the grounds for appeal. (ii) Deposit with the Metal Market & Exchange Co. Ltd. the sum of £200, and in addition the sum, if any, which shall be payable under the Award by the Appellant.

(b) Upon receipt of such Notice of Appeal the Committee shall within 4 weeks nominate not less than five members, (hereinafter called "the Appeal Committee") to hear the Appeal. Members of the Appeal Committee shall be members of the Committee of the London Metal Exchange and/or members of the Board of the Metal Market & Exchange Co. Ltd.

(c) The procedure on appeal shall as far as possible be similar to that above provided for the original hearing except that all statements and documents delivered to the Secretary under Rule 7 (l) shall be laid before the Appeal Committee who may, however, require such further statement or statements or other information or documents from either or both of the parties as the Appeal Committee may think necessary. The provisions of Rule 7 (k) shall apply in like manner to the deposit referred to in sub-paragraph (a) (ii) of this Rule as the deposit in connection with the original hearing.

(d) The decision in writing of the majority of the Appeal Committee (which latter shall not at any time number less than five) shall be final and binding on all parties, and the Appeal Committee shall also decide whether the whole or any part of the said deposit of £200 shall be returned to the Appellant or be forfeited.

(e) The Appeal Committee shall have the same discretion regarding costs as is given to the arbitrators under Section 7 (i) and shall fix the amount of their remuneration and direct by whom it shall be paid.

(f) All statements and all documents lodged with the Appeal Committee shall, together with the Award, be deposited by them with the Secretary by whom they shall be retained until the costs and fees specified in the Award have been paid by either party. On payment, which shall not affect any provision of the Award, a copy of the Award shall be delivered to both parties and all documents returned to the parties concerned.

Appendix 6:

Conversion table

Mass	Metric ton	Long ton	Short ton	Pounds	Kilograms	Ounce troy	Grains	Pennyweight	Pikul
Metric ton		0.984	1.102	2204.5	1000				16.535
Long ton	1.016		1.120	2240	1016				
Short ton	0.907	0.893		2000	907				15.000
Pound	0.0004	0.0004	0.0005		0.4536	14.583	7000		
Kilogram	0.001			2.2045		32.150			
Ounce troy							480	20	
Grain						480			
Pennyweight							24		
Pikul	0.0605		0.067	133.33	60.48				

Appendix 7:

Conversion of US metal prices from cents per lb to sterling pounds per metric ton* at varying exchange rates

Metal price: cents per lb	Exchange rate: US dollar/£ sterling									
	1.60	1.70	1.80	1.90	2.00	2.10	2.20	2.30	2.40	2.50
	£ sterling per metric ton									
1	13.77	12.96	12.24	11.60	11.02	10.49	10.01	9.58	9.18	8.81
5	68.85	64.80	61.20	58.00	55.10	52.45	50.05	47.91	45.91	44.08
10	137.70	129.60	122.40	116.00	110.20	104.90	100.10	95.82	91.83	88.16
20	275.40	259.20	244.80	232.00	220.40	209.80	200.20	191.64	183.66	176.32
30	413.10	388.80	367.20	348.00	330.60	314.70	300.30	287.46	275.49	264.48
40	550.80	518.40	489.60	464.00	440.80	419.60	400.40	383.28	367.32	352.64
50	688.50	648.00	612.00	580.00	551.00	524.50	500.50	479.10	459.15	440.80
60	826.20	777.60	734.40	696.00	661.20	629.40	600.60	574.92	550.98	528.96
70	963.90	907.20	856.80	812.00	771.40	734.30	700.70	670.74	642.81	617.12
80	1,101.60	1,036.80	979.20	928.00	881.60	839.20	800.80	766.56	734.64	705.28
90	1,239.30	1,166.40	1,101.60	1,044.00	991.80	944.10	900.90	862.38	826.47	793.44
100	1,377.00	1,296.00	1,224.00	1,160.00	1,102.00	1,049.00	1,001.00	958.20	918.30	881.60

**metric ton = 2,204 lb*

Appendix 8:

Conversion of tin prices from Malaysian ringgits per pikul* to sterling pounds per metric ton at varying exchange rates

Malaysian ringgits/pikul	Exchange rate: Malaysian ringgits/£ sterling 3.5	4.0	4.5	5.0
	£ sterling per metric ton			
500	2,361	2,066	1,836	1,653
1,000	4,722	4,132	3,673	3,306
2,000	9,444	8,264	7,346	6,612
3,000	14,166	12,396	11,019	9,918

**pikul = 133.33lb*